An Alternative

Theory of Global Warming and Cooling

By

George T. Croft, Ph.D.

1

TABLE of CONTENTS

Synopsis: An Alternate Theory of Global Warming and Cooling.

Newspapers, television, radio, and almost any form of mass communication has or continues to produce documentaries on the consequences of Global Warming. Yet there is controversy as to whether global warming is in fact occurring or increasing, and if so the cause.

It is generally accepted by geologist and other members of the scientific community that about 9,500 years ago the Northern Hemisphere was covered by thick ice sheets at latitudes above about $41°$ north latitude. This massive ice field, referred to as the last ice age began to retreat about 9,500 years before present time. Was the retreat of the ice sheets the result of 'Global Warming' and if so what caused the onset of warming? Is the warming simply a continuation of warming started 9,500 years BP[1] or is it caused by an increasing concentration of green house gases or is there yet another cause?

Since the Sun is the source of the ***energy received***[2] by Earth, something must have happened to this source or to the way it is distributed over Earth to cause the advance of the ice sheets and later, to cause the retreat of the ice sheets. Changes in energy received from the Sun by Earth are caused by changes in Earth's distance from the Sun, the spin axis tilt angle, and direction of Earth's axis of rotation (spin axis) relative to the orbital plane[3].

In this monograph a theory is developed based on well established knowledge and principles of physics and astronomy that correctly predicts the recent ice age, the warming observed BP and an imminent onset of global cooling. In addition, the theory developed correctly predicts, as one should expect, the seasons, and the reason hurricanes start near the Tropic of Cancer and Tropic of Capricorn, respectively. The role of so called "green house gases" in determining the temperature of Earth is also discussed in detail.

Equations, based on established physical laws, are derived and used to calculate the effects of changes in the eccentricity of Earth's orbit, the spin axis tilt angle, and the of the direction of the spin axis (the precession angle) on the energy received at any point on Earth.

The equations developed predict the current seasons, as they should, but also predict a quasi-periodic variation in energy received by Earth as the precession angle advances counterclockwise. Many graphs showing the energy received by Earth for a variety of precession angles and the orbital position of Earth are contained in the monograph.

Using the theory developed in this monograph the following results and predictions are obtained. They are;
1) The global warming is a natural continuation of the warming cycle that began 9,500 years BP. Hence, Earth has been in a warming cycle since the beginning of the retreat of the North American ice sheets.
2) The predicted rate of change of temperature over the last 600 years increases with latitude and has been between 0.06 to $0.65°C$ per century.
3) The predicted current rate of change of temperature is approaching zero and, if not already occurring, Earth will in the immediate future begin a cooling cycle that can ultimately cause onset of another ice age.
4) The theory predicts, correctly, the time and latitude of onset of the retreat of last ice sheet in North America.
5) After reaching saturation level of green house gas concentration, further increases in concentration do not increase the temperature.
6) Current activity to mitigate the effects of global warming by reducing green house gas concentration could be counter productive and accelerate the onset of global cooling.

George T. Croft, Ph.D. 22 Coventry Court, Bluffton, SC 29910, October 23, 2009

---

[1] BP is an abbreviation of before present time.
[2] **Energy received** by the Sun as used in the monograph, means the energy entering the upper most atmosphere.
[3] The intensity of cosmic rays emitted by the sun also follows an 11 year cycle. Cosmic rays can influence cloud formation and thereby cause a decrease in energy <u>incident</u> upon the surface of Earth.

Introduction: Global Warming and Cooling.

Since the 19th Century many theories that attribute warming of the air near the surface of Earth to water vapor and carbon dioxide have evolved[4]. Some of the energy radiated by the Sun reaches the surface of Earth where it is absorbed and then re-emitted as infrared radiation. So called 'Green House Gases' (water, H_2O, carbon dioxide CO_2, and, methane, CH_4) can absorb infrared radiation emitted by Earth thus increasing the energy of these molecules. This energy is transferred to the other gases in air thereby raising the temperature of the air near the surface. Thus it is reasonable to conclude that an increase in GHG causes an increase in temperature at Earth's surface and in its atmosphere[5].

Others have asserted that a warmer Earth will have more clouds that will reduce the energy received by Earth and thereby cause cooling that leads to an increase in the snow coverage that in turn reflects the incoming energy from the Sun back to space. Yet others think that short summers, for what ever reason, lead to more snow cover in the higher latitudes that in turn, reflects more energy back to space, thus causing build up of glaciers and global cooling. There are many theories supporting either global warming or cooling, including the Milankovitch theory that relates the glacial ages to changes in Earth's orbit, the precession of Earth's spin axis, and change of Earth's spin axis relative to the orbital plane.

One fact remains, namely, that all of the theories deal with energy received by and distributed on Earth. Hence, it is important to understand how Earth receives energy. Energy received by Earth comes predominantly from the Sun. Some energy also comes from within Earth intermittently from volcanic eruptions, but the energy from the later compared to that received directly from the Sun is small.

The energy arriving at the surface of Earth can and has been modulated by ***unpredictable events***. For example, particles in the atmosphere caused by a volcanic eruption or a comet or meteorite collision with Earth decrease the energy transmitted to surface of Earth.

Given that Earth's primary source of energy is the Sun, it becomes clear that we must understand just how that energy is distributed on Earth. The ***energy received***[6] by Earth is a function of the variables that describe Earth's orbit, tilt of the spin axis relative to the orbital

plane, Earth's spin axis precession, and, variations in the intensity of the energy produced by the Sun. Energy distribution and corresponding temperature distribution becomes the starting point for understanding the climate change on Earth.

In Chapter I, of this monograph all the equations and computer programs required to calculate and display the distribution of ***incoming energy*** per unit time (power) from the Sun over the surface of Earth as a function of the rotation of Earth about its spin axis, the precession of Earth's spin vector, the angle of the spin vector relative to the orbital plane, and, the orbital position of Earth are derived.

In Chapter II, a technique is developed that relates the **energy received**[7] to temperature. Researchers to date have concentrated on elaborate detailed models or combinations of several models to predict temperature. In this monograph, the author has used data obtained by direct measurement of zonal average temperature as a function of latitude for a year to relate temperature to calculated energy received. Using this approach the need

---

[4] Spencer Weart, The Discovery of Global Warming, http://aip.org/history/climate/simple.htm, 06/2008

[5] The nominal CO_2 concentration in air is about 0.04% or 400 per million air molecules. Assuming some monotonically increasing functional dependence of temperature on GHG concentration one can argue that increasing the GHG concentration causes an increase in temperature of the air.

[6] See footnote 2

[7] See footnote 2

for detailed assumptions in a complicated model is eliminated, and interpretation of results is simplified. Since the directly measured equilibrium temperature is the result of a balance between input energy and loses to space, detailed effects of the GHG in the atmosphere on energy retention or loss and Stephan-Boltzman radiation are accounted for with a single thermodynamic equation of state that relates the thermodynamic coordinates, surface air temperature and internal energy.

In Chapter III, data obtained by measuring temperature over a range of latitudes using different techniques is compared to predictions of the theory. Also an additional theory is suggested that clarifies the role of green house gases in temperature change.

Causes for Fundamental Changes in the Energy Received from the Sun.
Because the distance from the Sun and Earth changes, as does the angle of Earth's spin vector relative to the orbital plane and the precession angle relative to a fiducial direction, the energy received from the Sun changes. The change in total energy entering the outer most limits of Earth's atmosphere at a given latitude over the course of a year is small, but the change in energy received in the summer and the winter seasons is not insignificant.

Because Earth is spinning about the North/South Pole axis, Earth bulges at the Equator. This imbalance in mass leads to a clockwise torque applied to Earth by both the Moon and the Sun. The torque exerted by the Sun is about an axis in the orbital plane. As a result, the spin axis precesses about an axis normal to the orbital plane. Precession of Earth's spin axis causes significant seasonal changes in the distribution of the energy from the Sun over the surface of Earth. According to Milankovitch theory, precession and an accompanying change in eccentricity was the cause of the recent ice ages[8]. In addition, the re-distribution of ice and snow caused by weather can cause changes in the rate of precession[9]. Hence, an equation describing the distribution of energy received from the Sun over Earth as a function of;
1) *The rotation of Earth about its spin axis,*
2) *The precession of Earth's spin vector,*
3) *The angle of the spin vector relative to the orbital plane,*
4) *The latitude and longitude of every point on Earth, and,*
5) *The position of Earth in Orbit,*
is *fundamental* to gaining an understanding of global warming and cooling phenomena.

Unpredictable Events.
Unpredictable events such as massive volcanic eruptions, collisions with Earth by comets, and, large meteorites are known to cause global cooling for times as short as years and as long as millennia. These unpredictable events are superimposed on the fundamental changes, caused by celestial motion, that influence global warming and or cooling. Unpredictable events are not part of the following analysis, however from geological evidence we know when they have occurred.

Variation in the Intensity of the Sun.
It is well known that the intensity of the Sun depends on sun spot cycles with a 11 year cycle. sun spot activity changes the *energy radiated* by Sun about 0.1%. Hence, in this analysis, the intensity of the Sun is treated as a constant[10].

---

[8] J. Imbrie and K.P. Imbrie, Ice Ages, , Harvard Univ. Press, 1979, Chapters 8,9,10 ,11,12
[9] Any increase in snow or ice at either pole can cause an increase in the mass at either or both poles. Such an increase in mass leads to a slight increase in the counterclockwise torque applied by the Moon and Sun that in turn decreases the rate of precession of Earth's spin axis and increases the tilt.
[10] Sun spot activity is associated with an increase in cosmic ray intensity. Cloud cover is correlated with cosmic ray activity as is energy received at the surface of Earth.

The Overall Objective.

1) Develop the equations for the distribution of energy received from the Sun on Earth as a function of known celestial variables,
2) Develop a method of relating temperature at Earth's surface to the energy received at the beginning of the atmosphere.
3) Compare theoretical predictions with data obtained from direct measurement of the temperature and from geological studies.
4) Use the result to predict future temperature of Earth.

Chapter I Energy from the Sun at Any Point on Earth.

The primary source of the daily flow of energy to Earth is the Sun. The energy received at a point on the surface of Earth is a function of the latitude, L, the longitude, ϕ, the angle between Earth's spin axis and a normal to the orbital plane, γ, the angle of precession of the spin axis, ω, the eccentricity of Earth's orbit, ε, the angular displacement of Earth, θ, about the Sun, and the distance between Earth and the Sun, R. In addition to the variables related to the Earth's orbit and tilt, the energy output from the Sun changes with a period of eleven of years. The energy retained by Earth also depends on conditions within the atmosphere that change hourly and the emissivity of Earth at each point on the surface[11].

The equation that describes the energy received by Earth from the Sun as a function the variables, L, ϕ, γ, ω, θ, ε, and, R is derived below. Two Cartesian coordinate systems are used (See Figure (1)). The first coordinate system is the conventional earth coordinate system that uses latitude, L, and longitude, ϕ, to locate a point on Earth. The earth coordinates, referred to as the unprimed system, are x, y, and, z. The second coordinate system is the Sun coordinate system. The sun system is referred to as the primed system with the x', y' axes in the orbital plane and the z' axis normal to the orbital plane. The precession angle ω, is the angular displacement of the projection of the Earth's spin axis onto the orbital plane relative to the sun positive x' coordinate.

As the Earth and its coordinate system rotate around a line parallel to the Sun's z' axis, the z axis of Earth (the spin vector) sweeps out a cone. The spin axis of Earth is a vector in the same direction as the + z axis of Earth. One can think of the spin axis as a vector normal to the *equatorial plane of Earth*. The sunrays are essentially parallel to the *orbital plane*[12]. The power per unit area carried by a bundle of sunrays passes through a plane normal to the direction of the sunrays. If we let the direction of the rays be the unit vector[13], $\mathbf{i_s}$, and, S, be the power per unit area normal to the plane, the energy flowing in the direction of the bundle of sunrays is[14,15],

$$S\mathbf{i_s} = S[Cos(\theta)\mathbf{i'} + Sin(\theta)\mathbf{j'}],$$

..........eq.(1)

where, θ, is the angle the sunrays make with respect to the x' axis. The displacement, **R**, between the Earth and the Sun's center of mass is given by,

$$\mathbf{R} = |R(\theta)|[Cos(\theta)\mathbf{i'} + Sin(\theta)\mathbf{j'}].$$

.........eq.(2)

Thus the flow of energy from the Sun is always in the direction of the vector between Earth and the Sun[16]. Since the Sun is 93,000,000 miles from Earth, the sunrays arriving at Earth are essentially parallel one to another and parallel to the orbital plane. Hence, the incoming energy is treated as a plane wave.

The power, P_N, received by Earth at a given point located by the vector, **V**, is the component of the sunrays normal to Earth. Hence[17],

---

[11] The energy received in the calculations is averaged over of a year and does not include effects Sun spots on energy received.
[12] The Sun radiates energy in a direction normal to its surface. Energy intercepted by Earth is within a cone of radius 4,000 miles and 93,000,000 miles from base to the apex. The angle between the central ray and the outer most ray is only 9 seconds of arc.
[13] $\mathbf{i_s}$ is described in the Sun coordinate system.
[14] S is, approximately 1,400 watts/square meter at a fixed distance from the Sun, that is always just outside Earth's atmosphere. Once the radiation wave front enters the limit of the atmosphere some energy is absorbed or reflected back to space.
[15] "Sun." Encyclopedia Britannica, from Standard Edition, 2008
[16] This follows because the vector $\mathbf{i_s}$ is always parallel to **R** and therefore to the orbital plane.
[17] The vector dot product of two vectors is the projection of one vector on the other.

$$P_N = S\mathbf{i_s}*\mathbf{V}/|\mathbf{V}|,$$

where, $\mathbf{V}/|\mathbf{V}|$, is a unit vector normal to the surface of Earth at a specific latitude, L, and longitude, ϕ. In addition, the energy received depends on Earth's distance from the Sun, R, which is a function of θ. The steps involved in the explicit evaluation of the energy distribution as a function of the variables, L, ϕ, γ, ω, R, and, θ, follow.

Figure (1) shows the equatorial plane of Earth tilted at an angle, $\pi/2 - \gamma$, with respect to the orbital plane of Earth. The vector that locates a point on the surface of Earth is, $\mathbf{V}$. The coordinates of $\mathbf{V}$ are in the earth coordinate system and are along the mutually orthogonal x, y, and, z axes. The spin axis of Earth is along the z coordinate. The sun system of coordinates is fixed in space. The x' and y' axes lie in the orbital plane and the z' axis is normal to that plane. The z axis, (the North-South pole spin axis), rotates around a line parallel to the z' axis sweeping out a conical surface. This motion is called the precession of the equinoxes. The precession angle, ω, is zero if the projection of the spin vector on the x'y' plane is in the direction of the positive x' axis. In this analysis, the angle of precession, ω, is the counterclockwise angular displacement between the *projection* of the z axis on the x'y' plane and the positive x'axis. Currently, at Winter Solstice, the projection of the spin vector is away from the Sun and in the negative x' direction. If we describe the Earth's orbit with the equation,

$$R = a(1-\varepsilon^2)/(1-\varepsilon Cos(\theta)),$$

at Winter Solstice, $\theta=\pi$, and $\omega=\pi$ degrees. Note that Earth is closer to the Sun and the spin vector points away from the Sun. If $\theta=0$ and $\omega=\pi$ and the projection of the precession angle vector points toward the Sun it is Summer Solstice and Earth is farthest from the Sun.

Figure (2) shows the trajectory of Earth about the Sun. Earth moves counterclockwise around the Sun and is always in the x'y' plane. The trajectory of Earth is described in the Sun system of coordinates. The vector locating Earth relative to the center of mass of the Sun is $\mathbf{R}$. The angular displacement, θ, of the radius vector, $\mathbf{R}$, is measured counterclockwise relative to the x' axis. Since the spin angular momentum is constant, the spin vector does not change magnitude or direction as Earth rotates about the Sun. Actually, the direction of the spin axis does indeed change because of torque applied to the gyroscope Earth by the Sun and the Moon. This causes the spin axis to rotate counterclockwise completing one revolution in 26,500 years. In addition, the angle between the spin axis and the orbital plane decreases by about 2.5° in 26,500 years. Thus for time intervals short compared to 26,500 years, the spin axis is considered fixed in direction and magnitude.

The large planets in our universe perturb the earth-sun system. This perturbation causes the earth-sun system to rotate counter clockwise about the system center of mass. Corrected for this perturbation, the time to complete the counterclockwise revolution of the spin axis is 22,000 years.

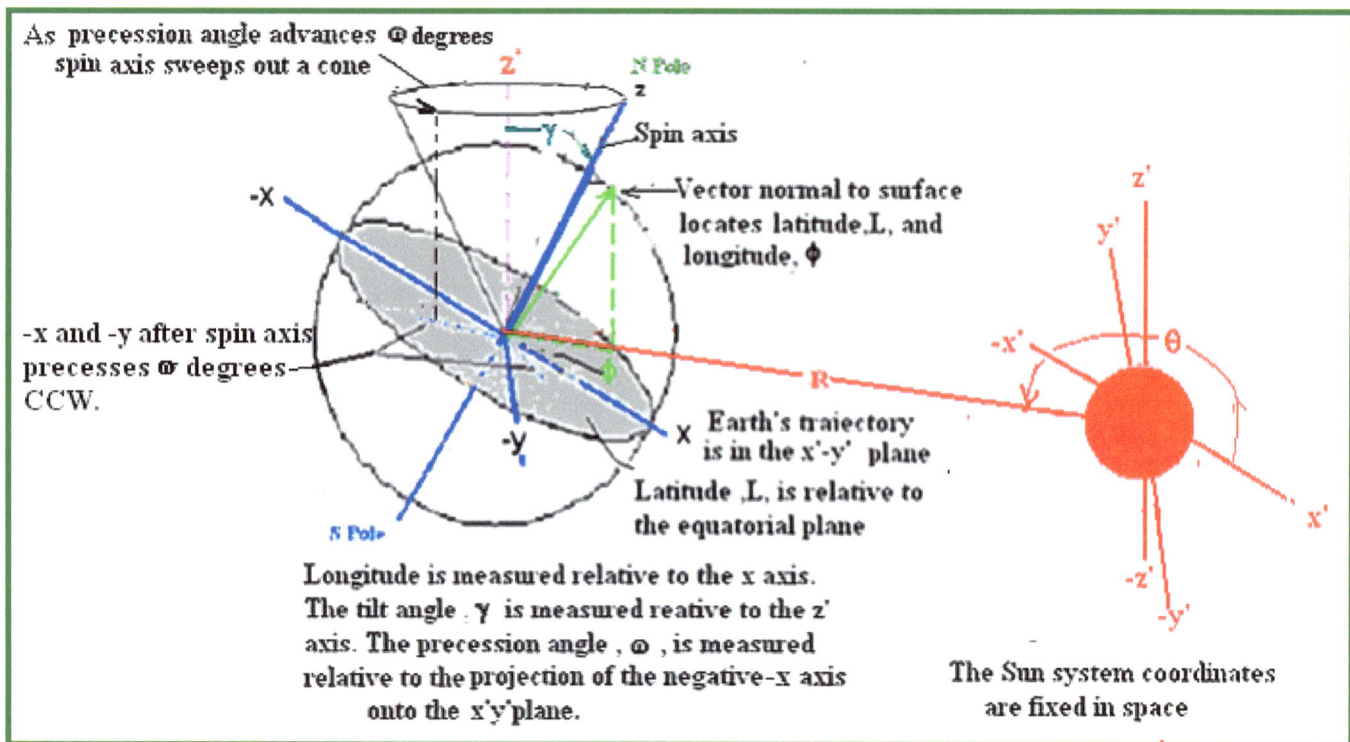

Figure (1) ↑

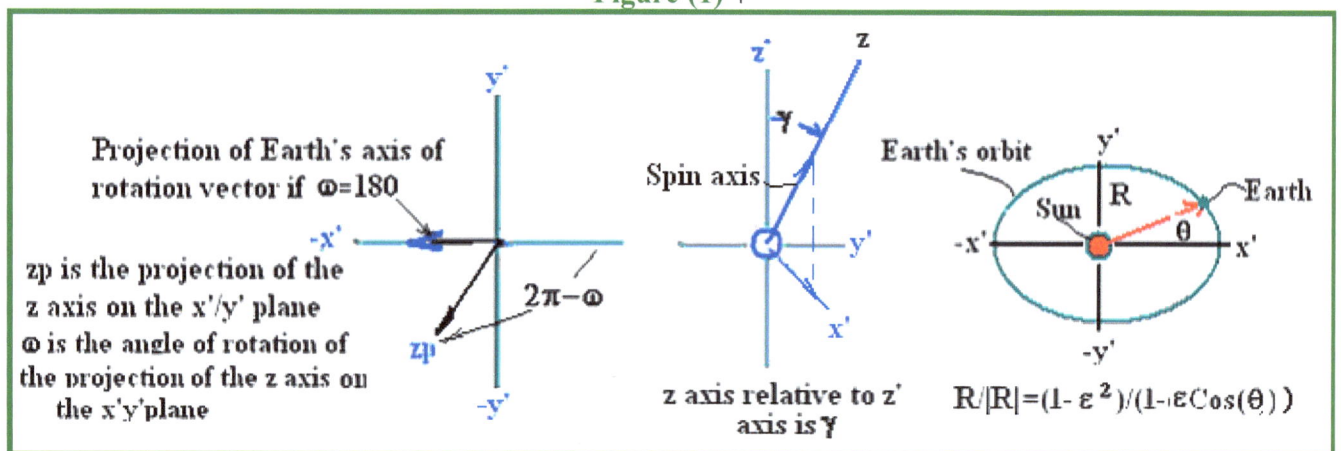

Figure (2) ↑　　　　　　　　Figure (2a) ↑　　　　　　　　Figure (3) ↑

The relationship between the earth and sun coordinate systems is shown in figures (1), (2), and, (3). In figure (1), as ω increases, the spin axis vector sweeps out the surface of a cone. If as shown in figure (2a), $\omega=0^o$, Earth's spin vector is in the x'z' plane and pointing in the positive x' direction. If $\theta=0^o$ and $\omega=180^o$, it is Summer Solstice. If $\theta=90^o$ and $\omega=180^o$ it is Fall Equinox. If $\theta=180^o$ and $\omega=180^o$ it is Winter Solstice. If $\theta=270^o$ and $\omega=180^o$ it is Spring Equinox. Earth's orbit about the Sun, see figure(3) is given by, $R/|R|=[|(1-\varepsilon^2)/(1-\varepsilon Cos(\theta))|]* [Cos(\theta)i'+ Sin(\theta)j']$ where, ε, is the eccentricity of Earth's orbit about the Sun. If $\omega=180^o$, the current configuration, Earth is nearest the Sun at Winter Solstice and farthest at Summer Solstice. If $\omega=180^o$ and $\theta=0^o$, the Earth is farthest from the Sun and it is Summer Solstice. If $\omega=180^o$ and $\theta=180^o$, the Earth is closest to the Sun and it is Winter Solstice. Note that if $\omega=0^o$ and $\theta=0^o$ it is colder in summer and winter, $\theta=180^o$, is warmer than for the current precession angle of $\omega=180^o$. Not only is Earth is *closer* to the Sun but also the spin axis points *toward* the Sun. Thus the position of Earth in its orbit combined with the orientation of the precession vector can create significant changes in energy received by Earth from the Sun.

Locating a Point on the Surface of Earth in the Sun Coordinate System. To obtain the energy falling on Earth, using equation (3), we must write the vector, **V(L,ϕ)**, which is in the Earth coordinate system in terms

10

of the sun coordinate system[18]. Since the magnitude and direction of, **V,** is independent of the coordinate system we can write,

$$V'=V.$$

Thus,

$$V'=x'\mathbf{i}'+y'\mathbf{j}'+z'\mathbf{k}'=|V|[Cos(L)Cos(\phi)\mathbf{i}+ Cos(L)Sin(\phi)\mathbf{j}+ Sin(L)\mathbf{k}]…..eq.(4)$$

and,

$$x= (|V|)Cos(L)Cos(\phi),\ y=(|V|)Cos(L)Sin(\phi),\ and,\ z=(|V|)Sin(L).$$

L, is the latitude of the point and, ϕ, is the angular displacement of the point from the Earth, **i**, axis. The primed unit vectors are in the sun system of coordinates, and the unprimed unit vectors are in the earth system of coordinates.

Multiplying equation (4) by **i', j'** and **k'** we obtain[19],

$$x'=[|V|][((Cos(L)Cos(\phi))\mathbf{i}*\mathbf{i}'+(Cos(L)Sin(\phi))\mathbf{j}*\mathbf{i}'+(Sin(L))\mathbf{k}*\mathbf{i}']$$
$$y' =[|V|][(Cos(L)Cos(\phi))\mathbf{i}*\mathbf{j}'+(Cos(L)Sin(\phi))\mathbf{j}*\mathbf{j}'+(Sin(L))\mathbf{k}*\mathbf{j}']$$
$$z'= [|V|][(Cos(L)Cos(\phi))\mathbf{i}*\mathbf{k}'+(Cos(L)Sin(\phi))\mathbf{j}*\mathbf{k}'+(Sin(L))\mathbf{k}*\mathbf{k}']$$

....eqs.(5a)

Thus,

$$V'=(|V|)\{[(Cos(L)Cos(\phi))\mathbf{i}*\mathbf{i}'+(Cos(L)Sin(\phi))\mathbf{j}*\mathbf{i}'+(Sin(L))\mathbf{k}*\mathbf{i}']\mathbf{i}'+$$
$$[Cos(L)Cos(\phi))\mathbf{i}*\mathbf{j}'+(Cos(L)Sin(\phi))\mathbf{j}*\mathbf{j}'+(Sin(L))\mathbf{k}*\mathbf{j}']\mathbf{j}'+$$
$$[Cos(L)Cos(\phi))\mathbf{i}*\mathbf{k}'+(Cos(L)Sin(\phi))\mathbf{j}*\mathbf{k}'+(Sin(L))\mathbf{k}*\mathbf{k}']\mathbf{k}'\}$$

....eqs.(5b)

Equation (5a) represents the components in the sun coordinate system of the vector, V'. The vector V' (equation (5b)) locates a point on the surface of Earth written in terms of the sun Coordinate system. The vector, V' , is always normal to the surface of the Earth as is, V, because it is identical in magnitude a direction to V. V' locates, in the sun system of coordinates, a specific point on Earth located at latitude, L, and longitude, ϕ.

Thus we see that we to obtain x', y', and, z', (the components of the vector **V'** along the i', j', and, z' coordinates, respectively) we must evaluate the terms **i*i', i*i', k*i', i*j', j*j', k*j', i*k', j*k', and, k*k'.** Clearly, the dot products above are simply the projections of the earth unit vectors onto the i', j', and, k' unit vectors of the sun system coordinates.

Explicit Evaluation of the Dot Products of Earth and Sun Unit Vectors.

To obtain the projections of the earth unit vectors onto the sun system unit vectors one must find a sequence of rotations of the earth system of coordinates, relative to the sun system coordinates, that orient the earth system such that it that coincides with the known motion of Earth relative to the sun system. The direction of the equatorial plane of Earth is always parallel to the spin axis vector and the earth system z coordinate[20]. The spin axis vector is always normal to the equatorial plane and precesses counterclockwise (CCW) around a line parallel[21] to the z' sun system coordinate. Thus we must find a sequence of rotations of the earth coordinate system that orient the earth system of coordinates so that the z coordinate is always at γ degrees relative to the sun system z' coordinate and that allows the z coordinate (the spin axis vector) to processes around a line always parallel to the z', coordinate.

---

[18] To form the dot product, $S\mathbf{i}_s*V/|V|$, the vectors, **V** and $\mathbf{i}_s$, must be written in the same coordinate system. Thus **V** must be written in terms of the primed system of coordinates.

[19] The unit vectors **i', j'**, and, **k'** are a mutually orthogonal set as are the vector, **i, j,** and, **k.**

[20] A plane is represented by a vector normal the plane.

[21] A line parallel to the z' coordinate is chosen because Earth follows a trajectory in the orbital plane.

11

First, we orient the earth system of coordinates so that the z axis is at an angle, γ, relative to and can precess about the z' axis. Then we find the projection of the, **i**, **j**, and, **k**, unit vectors of Earth's coordinate system onto the, **i'**, **j'**, and, **k'** unit vectors of the Sun's coordinates.

To simplify visualization of the processes, we translate the earth system so that it is coincident with the sun system unit vectors[22]. To obtain the requisite orientation of the earth system relative to the sun system of coordinates we start with Earth's and Sun's unit vectors coincident. Next we rotate earth system, ω, degrees counterclockwise about the **k'** axis of the sun system. The earth system is now at a new orientation labeled the double primed system. Next we rotate clockwise about the new earth j coordinate, γ, degrees.

The resulting configuration leaves the Earth's spin axis (vector **k**) at an angle, γ, relative to the **k'** sun system unit vector and the equatorial plane of Earth containing the **i** and **j** unit vectors at the desired tilt relative to the plane defined by the **i'** and **j'** unit vectors the sun coordinate system.

The Unit Vector Projections.
By inspection of the final configuration one can determine the projection of earth vectors onto sun vector. They are,

$$i*i'=Cos(\gamma)Cos(\omega), \quad j*i'=-Sin(\omega), \quad k*i'=Sin(\gamma)Cos(\omega),$$
$$i*j'=Cos(\gamma)Sin(\omega), \quad j*j'=Cos(\omega), \quad k*j'=Sin(\gamma)Sin(\omega),$$
$$i*k'=-Sin(\gamma), \quad j*k'=0, \quad k*k'=Cos(\gamma)$$

Thus **after** substituting in equation (5b) we obtain,

$$\mathbf{V'}=(|V|)\{[Cos(L)Cos(f)Cos(g)Cos(w)-Cos(L)Sin(f)Sin(w)+Sin(L)Sin(g)Cos(w)]\mathbf{i'}+$$
$$[Cos(L)Cos(f)Sin(w)Cos(g)+ Cos(L)Sin(f)Cos(w)+ Sin(L) Sin(w)Sin(g)]\mathbf{j'}$$
$$[-Cos(L)Cos(f)Sin(g)+ Sin(L) Cos(g)]\mathbf{k'}\}.$$
$$\dots\dots\dots\dots\dots eq.(5b)$$

If we let $L=\pi/2$ then V' is in the direction of the spin vector and,

$$V' = Sin(g)Cos(w)i'+Sin(g)Sin(w)j' + Cos(g)k',$$

showing that as the spin vector precesses it rotates CCW around the z' axis sweeping out a cone. Further, if w=0 the spin axis projection onto the x' axis is in the positive direction. If $w=\pi$ the projection is in the negative direction. Thus showing that equation (5b) is consistent with the sign convention as shown in figures (2), (2a), and, (3).

Using equation (3), the power per unit area, P_N, at a point on Earth located by the vector **V'** in the sun system of coordinates is,

$$P_N=S(V/|V|*i_s)=S(V/|V|)\{[Cos(L)Cos(f)Cos(g)Cos(w)-Cos(L)Sin(f)Sin(w)+ Sin(L)Sin(g)Cos(w)]Cos(\theta)+$$

$$[Cos(L)Cos(f)Sin(w)Cos(g)+ Cos(L)Sin(f)Cos(w)+ Sin(L) Sin(w)Sin(g)]Sin(\theta)\}$$
$$\dots eq(6)$$

Note that the term, z'**k'**, is not used because the Sun's rays are parallel to the x'y' plane and therefore have no **k'** component. After normalizing the equations by setting $|V|=1$, the equations describing the power per unit area on earth in the sun coordinate system are,

---

[22] If a vector is translated, the direction and sense of direction of the vector remains unchanged.

$$P_N=S\{[Cos(L)Cos(\phi)Cos(g)Cos(\omega)-Cos(L)Sin(\phi)Sin(\omega)+ Sin(L)Sin(\gamma)Cos(\omega)]Cos(\theta)+$$

$$[Cos(L)Cos(\phi)Sin(\omega)Cos(\gamma)+ Cos(L)Sin(\phi)Cos(\omega)+ Sin(L) Sin(\omega)Sin(\gamma)]Sin(\theta)$$

....eq.(7)

Equation (6) describes the normalized power per unit area at latitude, L, and longitude, ϕ, for a given value of γ (spin axis tilt angle) and ω (angle of precession).

Equation (7) converted to VB6 code becomes,

$$P_N=S*((Cos(L)*Cos(f)*Cos(g)*Cos(w)-Cos(L)*Sin(f)*Sin(w)+ Sin(L)*Sin(g)*Cos(w))*Cos(q)+$$

$$(Cos(L)*Cos(f)*Sin(w)*Cos(g)+ Cos(L)*Sin(f)*Cos(w)+ Sin(L)* Sin(w)*Sin(g))*Sin(q))$$

......eq.(7)

Correction for Changes in Earth-Sun Distance.
Earth's orbit about the Sun is an ellipse[23] and is described by the equation[24],

$$R=a(1-\epsilon^2)/(1-\epsilon cos(\theta)),$$

...eq. (8)

where R is the distance from the center of mass of the Sun[25] to the center of mass of Earth and, ϵ, and, θ, are the eccentricity of the orbit and angular displacement of the radius vector measured CCW from the postive x' coordinate of the sun system, respectively.

The energy received depends on the reciprocal of the square of the distance from the Sun. Hence, the energy must be corrected by the term,

$$1/R^2= 1/[a(1-\epsilon^2)/(1- \epsilon cos(\theta))]^2.$$

......eq.(9)

Normalizing the radius vector by dividing by the major axis, a, we obtain,

$$R/a=(1-\epsilon^2)/(1- \epsilon Cos(\theta))$$

hence,

$$1/(R/a)^2=(1- \epsilon Cos(\theta))^2/(1- \epsilon^2)^2$$

Since the power reaching Earth is proportional to the reciprocal squared of the Earth/Sun distance the corrected power becomes,

$$[1/(R/a)^2]P_N$$

..................eq.(10)

The corrected equation is,

$$P_{NR}(L,\phi,\omega,\gamma,\theta, \epsilon)= P_N(1- \epsilon Cos(\theta))^2/(1- \epsilon^2)^2$$

....eq.(11)

Equation for the Power from the Sun.
$P_{NR}(L,\phi,\omega,\gamma,\theta, \epsilon)$ *is the normalized power per unit flowing from the Sun to Earth corrected for Earth's position in orbit and for eccentricity of the orbit.* Converted for VB6 code and used in program **envslfgwcor.vbp** the equations become,

$$\mathbf{P_{NR}}= P_N*(1-z*Cos(q)^{\wedge}2/(1-z^{\wedge}2)^{\wedge}2$$

..................eq. (12)

---

[23] George T. Croft, Ph.D. Three Dimensional Analytic Geometry, Amazon.Com, 2000, Chapter IV,

[24] If $\theta=0$ the Earth is to the right of the origin where the Sun is located.

[25] Actually, the Earth revolves around the center of mass of the Sun/Earth system. Since the mass of the Sun is 3.32×10^5 times larger than that of the Earth, the center of mass of the Sun/Earth system is displace about 250 miles from the center of the Sun. Locating the radius vector at the center of the Sun introduces an error of $2.7 \times 10^{-4}\%$ in the Earth/Sun distance, which is negligible.

Power Received at Summer and Winter Solstice and the Equinoxes.

The equations (11) and (12) are used in VB-6, program[27] **envsLgwcor.vbp** to generate figures (5) through (15). The green circle represents the orbit of Earth about the Sun. The blue dot shows the position of Earth in orbit. The blue line represents the spin vector and the blue dot indicates the direction of the spin vector[28]. The doted lines are the sun system coordinates. Positive x' is toward the right (θ=0), and positive y' is toward the top (θ=90) of the picture. The column of numbers on the right list the latitude, and, in some cases, the power received at a specific value of rotation, ϕ, of Earth about the spin axis.

Power Received by Earth in a Day.

Graphs derived from equation (12) show normalized power received by Earth during the course of a day at Summer and Winter Solstice, and Fall and Spring Equinox. In addition, graphs are also shown that illustrate how the power received is influenced by the precession angle direction and the angle Earth's spin axis makes with respect to a normal to the orbital plane of Earth.

The graphs also show how the power received depends on both the latitude and the angle Earth's spin axis makes with respect to a normal to the orbital plane of Earth as well as the direction of the precession angle. The way power from the Sun is distributed on Earth essentially determines both the motion of the atmosphere, the ocean currents, glacier growth and retreat.

The influence these factors have on hurricane formation and path of travel, and the prevailing winds is also illustrated.

The Greek symbols used in the equations are replaced in the graphs by the corresponding English characters. For example, the Greek symbol gamma, γ,(the tilt angle) is replaced by the English character, g, the symbol, ω,(precession angle) is replaced by w and the symbol, θ,(angular displacement of Earth in its orbit) is replaced by q.

In figure (5), we see the power received by Earth at the Summer Solstice position. The length of a day increases as latitude increases. The energy received in a day is the integral of the power above the PNR=0 line. At the Equator, power flows to Earth from ϕ=90° to 270° or 12 hours. At 50° latitude power flows to Earth from ϕ=58° to 301° or 16 hours and 12 minutes. Above 66.55° Earth receives energy for 24 hours. At higher latitudes, the Sun never sets but does rise and fall relative to the horizon. Note that the maximum power/unit area, at approximately between 20 and 30 degrees latitude, is greater than that at the Equator. The power/unit area is maximum at L=23.45°. Maximum power/unit area for any given latitude is at high noon. The angle of elevation of the Sun above the horizon at high noon is equal to the angle whose cosine is $PNR/(1-\varepsilon Cos(q))^2$. Even though Earth is farthest from the Sun at Summer Solstice, the maximum power received is greater than at Winter Solstice because the spin axis leans toward the Sun in summer and away from the Sun in winter.

Figure (5), below, shows power received is plotted versus , ϕ, (angle of rotation about the spin axis) and, color coded latitude, L. The quantities, q, w, g, and, zo are, the angular displacement of Earth's radius vector from the apogee, the precession angle, the tilt angle, and, the eccentricity, respectively. Day time is above the PNR=0 line in the Northern hemisphere.

---

[27] Greek symbols are replaced by English characters because Visual Basic does not recognize Greek symbols.
[28] The direction of the spin vector changes approximately 0.014° per year (~50 seconds/year).

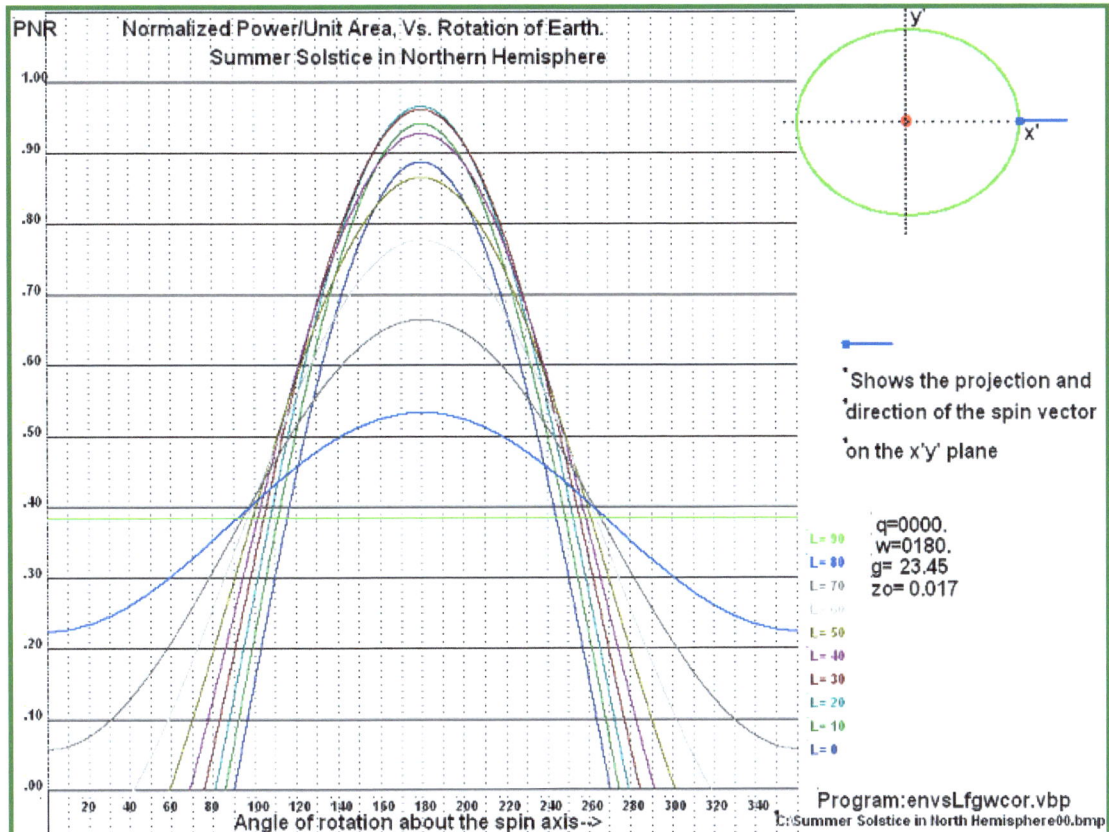

Figure (5)

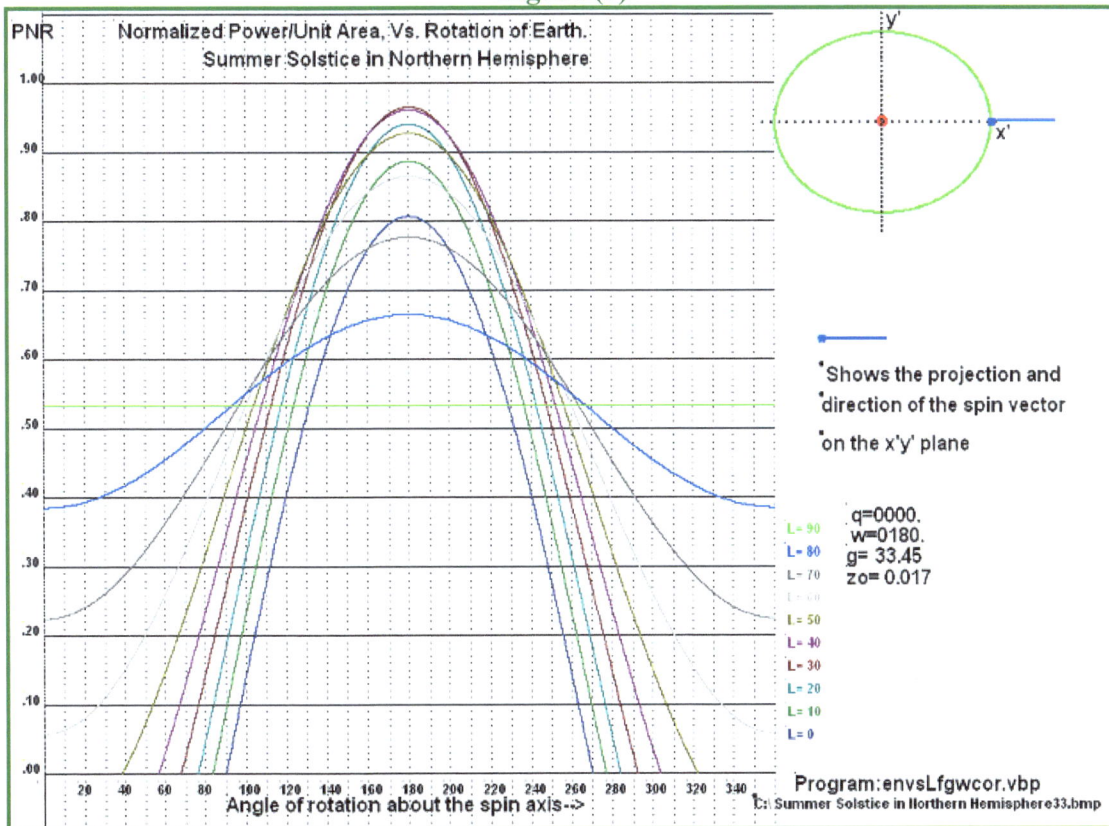

Figure (6)

Figure (6), above, shows the power received per unit area with the tilt angle, γ, increased 10° to 33.45^o. The maximum power shifts to between 30 and 40 degrees and is at exactly $L=33.45^o$. Thus showing that the maximum

15

Figures (5) and (6) clearly show that maximum power/unit area received depends on latitude. The theory predicts, correctly, that the current maximum power per unit area received is at L=23.45⁰ which is the latitude of the Tropic of Cancer in the Northern hemisphere. At high noon the Sun is directly overhead at the Tropic of Cancer.

Distribution of Energy Received.

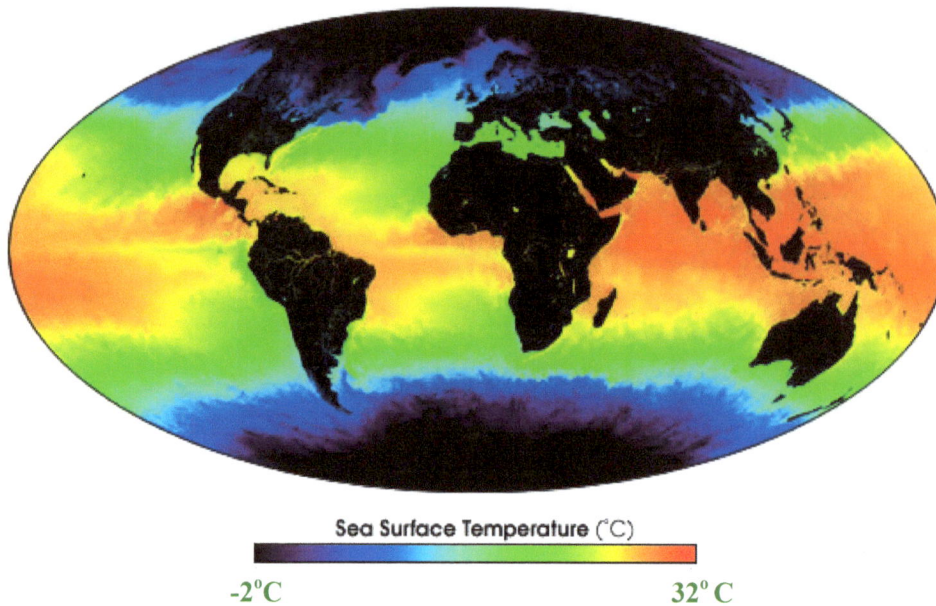

Sea Surface Temperature (°C)

-2⁰C   32⁰ C

Figure (7)

The figure shows the temperature around the world as measured between 05/01/01 and 05/30/01 by NASA. Since Summer Solstice is at approximately 06/31/01 the map represents the temperature distribution close to Summer Solstice. Note that for regions located at latitudes and longitudes entirely over ocean, the power delivered to regions near or on the Equator is less than at latitudes above and below the Equator.

The temperature distribution in figure (7) is as one would expect given the theoretical power shown in figures (6), (8), (9), and, (10). The theoretical maximum power per unit area from the Sun in the Atlantic and Pacific Oceans currently ranges between 23⁰ to 20⁰ latitude degrees above or below the Equator. Since the equilibrium temperature is controlled by the rate that power flows into Earth minus the rate that it flows from Earth, the temperature is highest where the input rate is highest as should be expected from the theoretical result in figures (6),(9), and (10

In figure (5), peak power develops in the Northern hemisphere at noon when the Sun is directly overhead at the Tropic of Cancer. Similarly, peak power develops in the Southern Hemisphere at noon when the Sun is directly overhead at the Tropic of Capricorn. The power per unit area flowing into the oceans is also at a maximum. Thus a temperature gradient from the vicinity of the Tropic of Cancer toward the North in the Northern hemisphere as well as a temperature gradient from the Tropic of Cancer toward the Equator is established. The later temperature gradient is opposed by yet another from the Tropic of Capricorn toward the Equator[29].

The temperature gradient starting at approximately 23.45⁰ latitude and extending toward the higher latitudes causes air and water to move toward the higher latitudes. The motion of the air and water is generally near or slightly north of the Tropic of Cancer toward the lower latitudes and the North Pole. This motion of air is not a simple uniform drift, but consists of many cells of circulating air. The prevailing upper atmosphere winds

---

[29] Temperature maps of the world recorded by NASA.

16

carry water vapor to the higher latitudes where the water vapor, depending on the local temperature, precipitates as rain and or snow. Having lost heat, the cold more dense air returns southward to replace the less dense rising warm air near the Equator(See figure(8)). Thus, circulation cells are formed that are distributed around Earth between a lower latitude of approximately 30° North and the North Pole region. Referring to figures (5), (9), and, (10) we see that as Earth's position advances beyond Solstice the maximum power received is at latitudes 2 to 4 degrees less than the Summer Solstice maximum latitude. This leads to a distribution of the maximum power received between L=23 to 20° clearly seen in figure (7), and predicted theoretically as shown in figures (9), and (10).

The circulation patterns below 30° toward the Equator are caused by the horizontal temperature gradients shown in figures (5), (6), (9), and, (10). In addition, circulation patterns are caused by the temperature gradients between the Tropic of Cancer and slightly cooler temperatures toward the Equator.

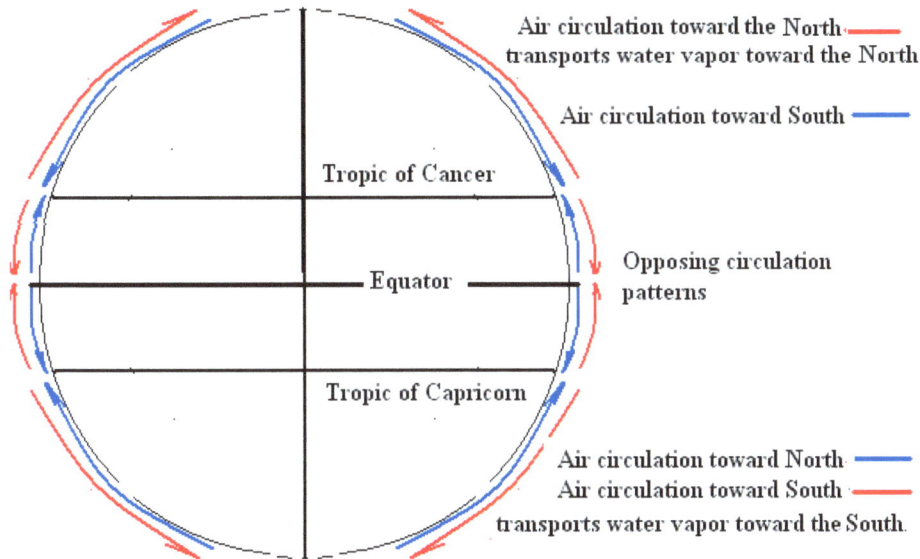

Figure (8)
Shows air circulation patterns that cause water to be transported from Tropic of Cancer and Capricorn to the higher latitudes. The over all effect of these patterns is transport of water or snow to the higher latitudes.

In figures (9) and (10), Earth is one month prior to and one month after Solstice. Note that even though the spin vector points slightly away from the line connecting Earth and the Sun, the power/unit area is slightly larger than at Summer Solstice because Earth is closer to the Sun than at Solstice.

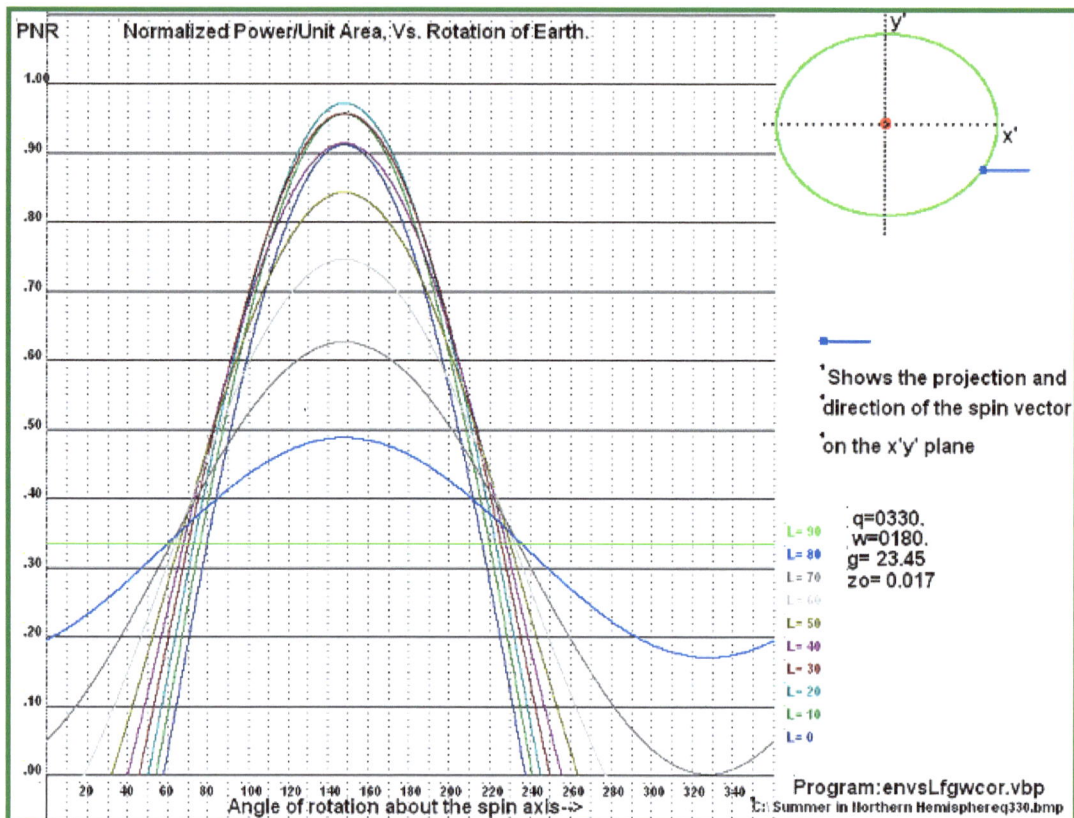

Figure (9)

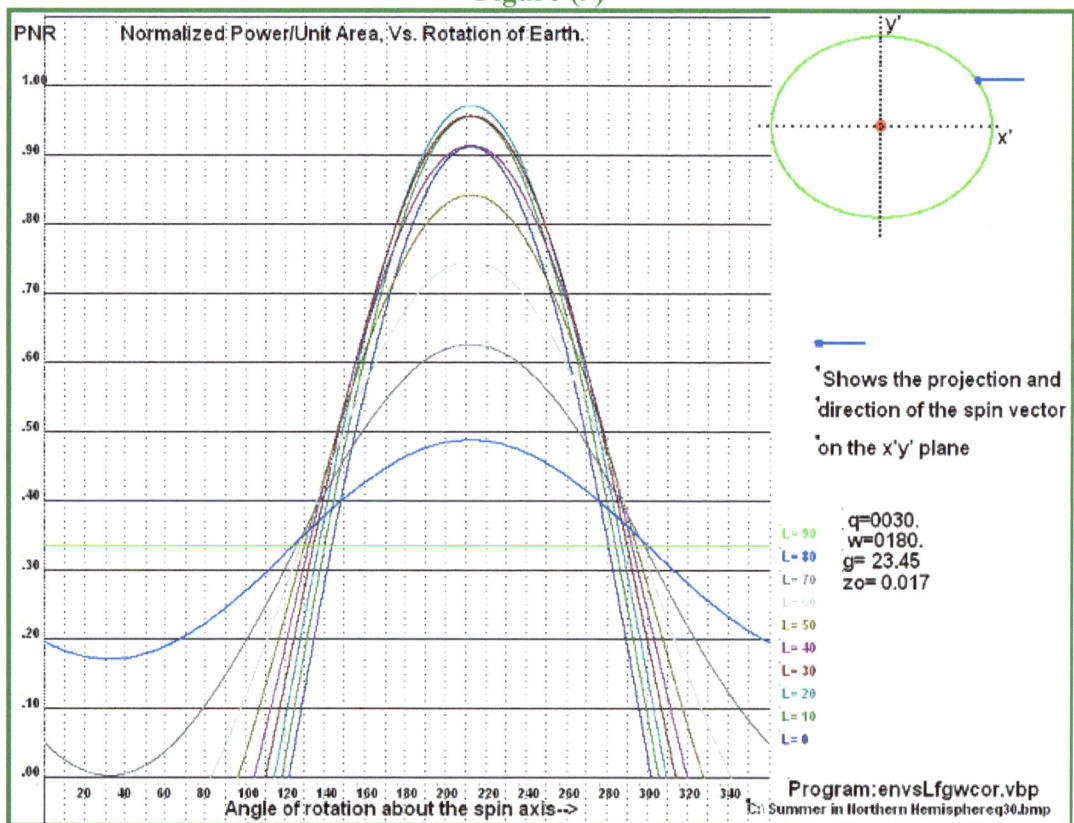

Figure (10)

Figure (9), (top), shows the power received/unit area one month before Summer Solstice. Maximum power occurs earlier than at Summer Solstice and is slightly greater than at Solstice because Earth is slightly closer to the Sun in May than June. The maximum is at $\phi=146^o$ and earlier than at Solstice. The normalized power received/unit area equals 0.9707 normalized units and is at L=20^0 when Earth is closer to the Equator than at Solstice.

Figure (10) (bottom) shows the power received/unit area one month after Summer Solstice. Maximum power/unit area received occurs later than at Summer Solstice and is slightly greater than at Solstice because Earth is closer the Sun in July than June. The maximum is at ϕ=212^0 and later than at Solstice. The normalized power/unit area equals 0.9707 units and is at L=20^0 and Earth is closer to the Equator. Thus the temperature gradient previously established between the Equator and maximum power received latitude (L=23.45^0) has now shifted to between the Equator and lower latitude (L=20^0). Yet another temperature gradient, in the opposite direction, is established between this broad temperature maximum gradient and points north of the Equator. This gradient provides the driving force that moves atmosphere toward the north in the Northern Hemisphere. This shift and spread of the latitudes for maximum temperature is seen in figure (7).

Horizontal and Vertical Temperature Gradients and Hurricanes.

Figure (11) Figure(12)

Hurricane Bob
Dates: 8/16 - 8/29 1991

Tropical Storm Erika
Dates: 9/8 - 9/12 1991

Hurricane Claudette
Dates: 9/4 - 9/14 1991

Figure (13)

Figures (11), (12), and, (13), show the points of origin and path of three recent Atlantic hurricanes[30]. All three originate 2 to 4 degrees latitude north of the Tropic of Cancer and drift toward the North.

Referring to figures (5), (6), (9), and, (10) we see that the Atlantic Ocean receives maximum power from the Sun at high noon near the Tropic of Cancer at Summer Solstice and that a horizontal gradient in power is towards North. The ocean surface temperature drives the rising heated air towards the North whilst Earth moves eastward. The result of these two motions is a clockwise drift of Atlantic Ocean hurricanes. A similar ocean current, for example the Gulf Stream, is also generated in part by temperature gradients in the ocean driven by energy received from the Sun.

---

[30] from Weather Underground, http://www.wunderground.com/hurricane

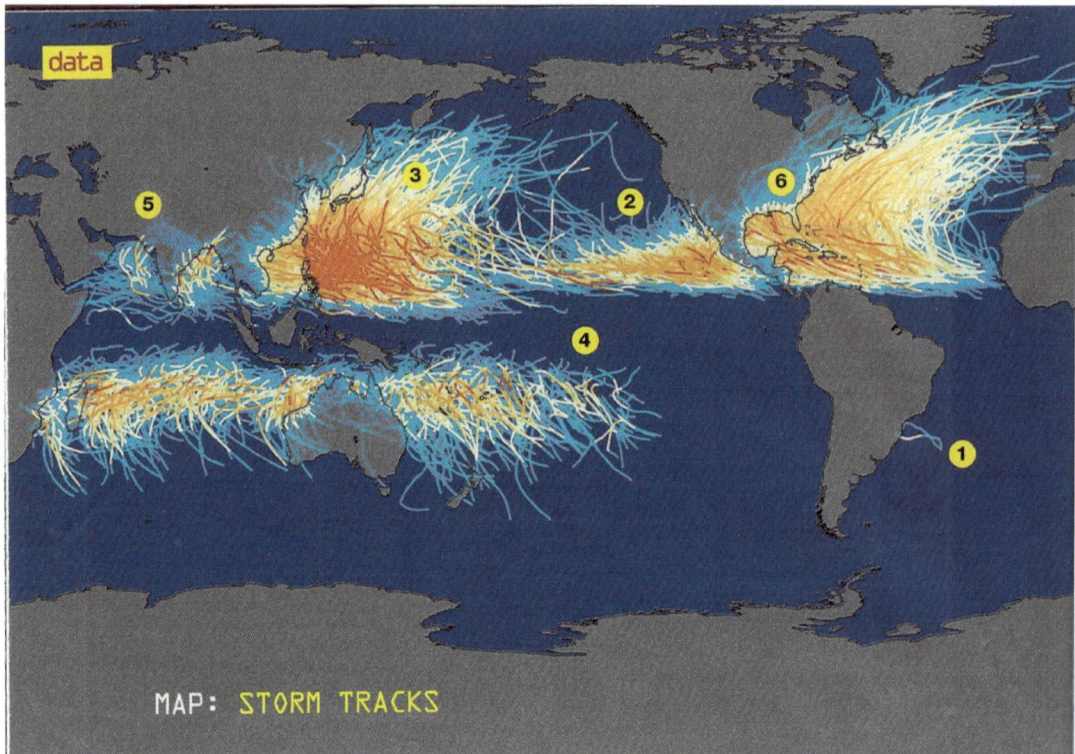

Figure (14)

Figure (14) shows the tracks of all _recorded_ hurricanes for the past 150 years[31]. Note that as described above, all North Atlantic and North Pacific hurricanes follow a clockwise path originating at or near the Tropic of Cancer or Capricorn. Hurricanes are observed in the summer months and start at latitudes below that of the Tropic of Cancer. However, all South Atlantic and South Pacific hurricanes follow a counterclockwise path. Since there is less land mass in the South Atlantic to intercept the trajectory of the hurricane less damage is reported and they are of less concern than their counter parts in the North Atlantic.

The temperature gradient between the latitudes near the Tropic of Cancer and further north latitudes persist as Earth travels around the Sun.

Figure (15) (See page 21)) shows PNR versus rotation of Earth about the spin axis at Autumn Equinox. The length of a day in both the Northern and Southern hemisphere is exactly 12 hours for all latitudes because the Sun's rays are normal to the spin axis at all latitudes. The North Pole (L=90) receives no power. The power received at high noon at the Equator at Autumn Equinox is greater than that received at the Equator at Summer Solstice. However, figures (15), (16), and, (17) show that the gradient in power received is maintained for all positions of Earth as it travels around the Sun.

---

[31] Robert Rohde, Map: Storm Tracks, Discover, September 2007, page 22

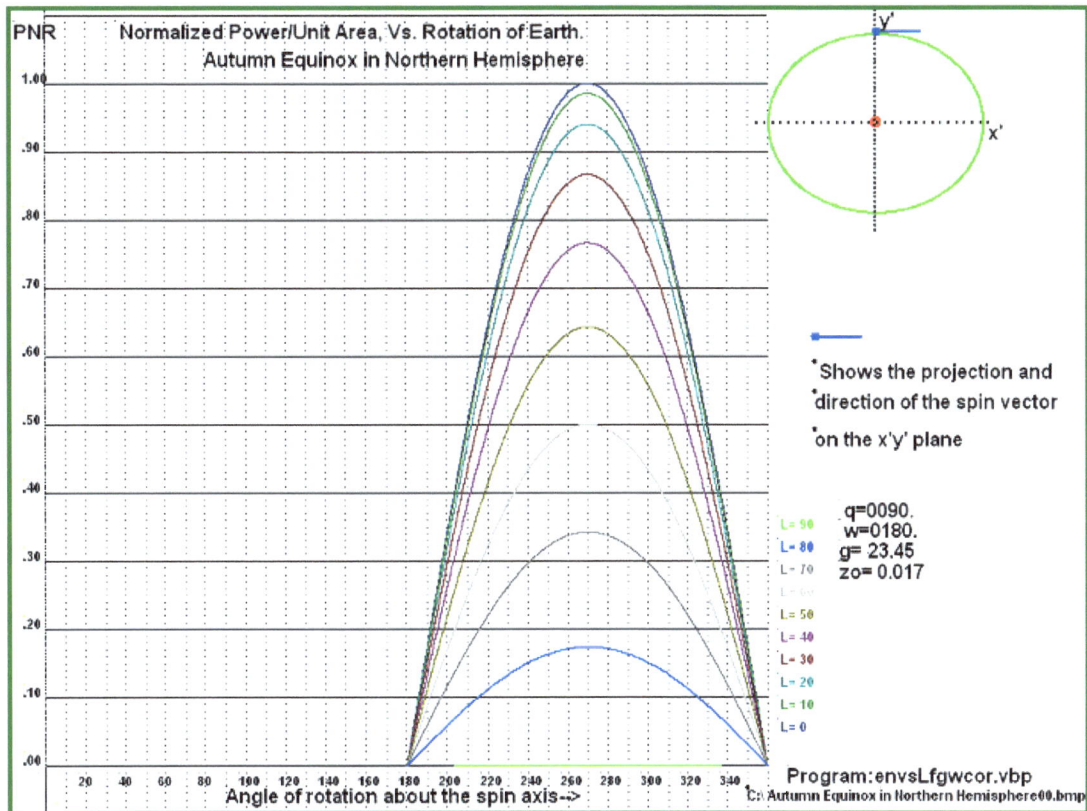

Figure (15)

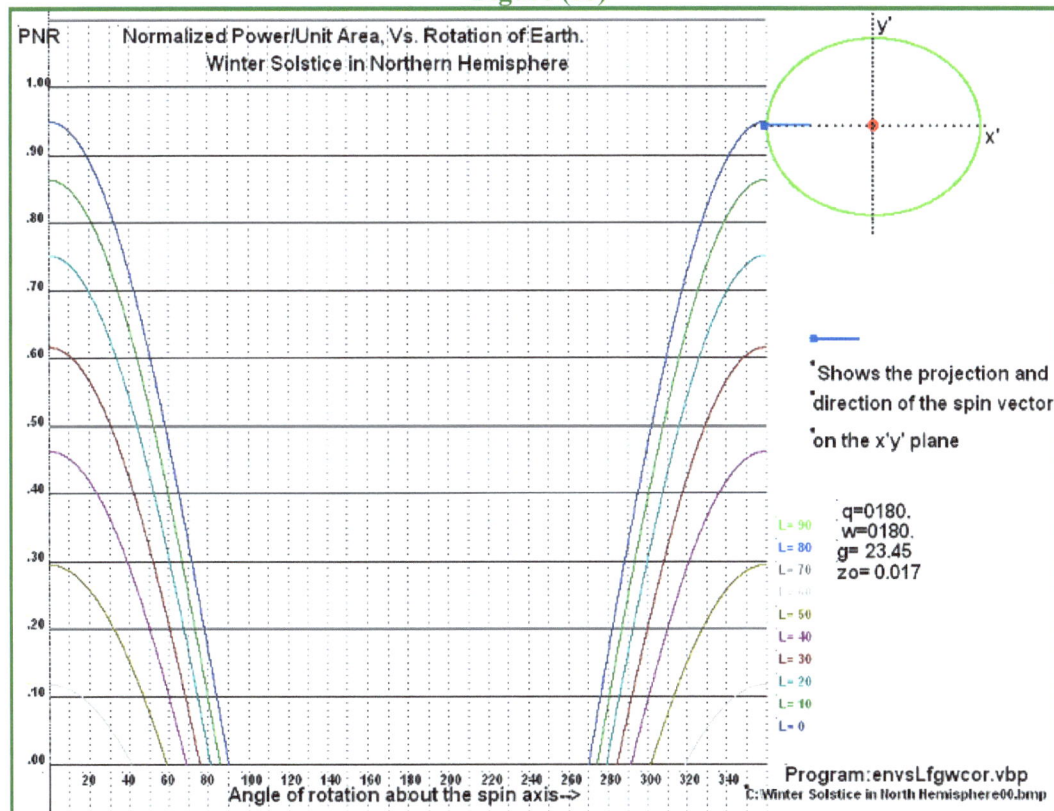

Figure (16)

Figure (16), above, shows power received at Winter Solstice as a function of the rotation of Earth about the spin axis and latitude. At high noon Earth is nearer the Sun than at high noon at Summer Solstice. Since the spin vector points away from the Sun, Earth receives less power than received at Summer Solstice. At latitudes above 66.45[0], the Sun rays do not strike Earth, and it is dark or nearly dark for 6 months. At 30⁰ latitude the length of a day is (150)*12 =10 hours 4 minutes. At 60⁰ latitude the day is only 84/180*12=5 hours and 36 minutes. Conversely the night at 60⁰ is 19 hours and 24 minutes. The day at the Equator is exactly 12 hours as it is for all seasons. If PNR is less than zero, Earth receives no power from the Sun and energy radiates power back into space. Hence, at night Earth's temperature declines, but as soon as day begins and power flows again from the Sun, Earth begins to warm. The details of this process are discussed in the next chapter.

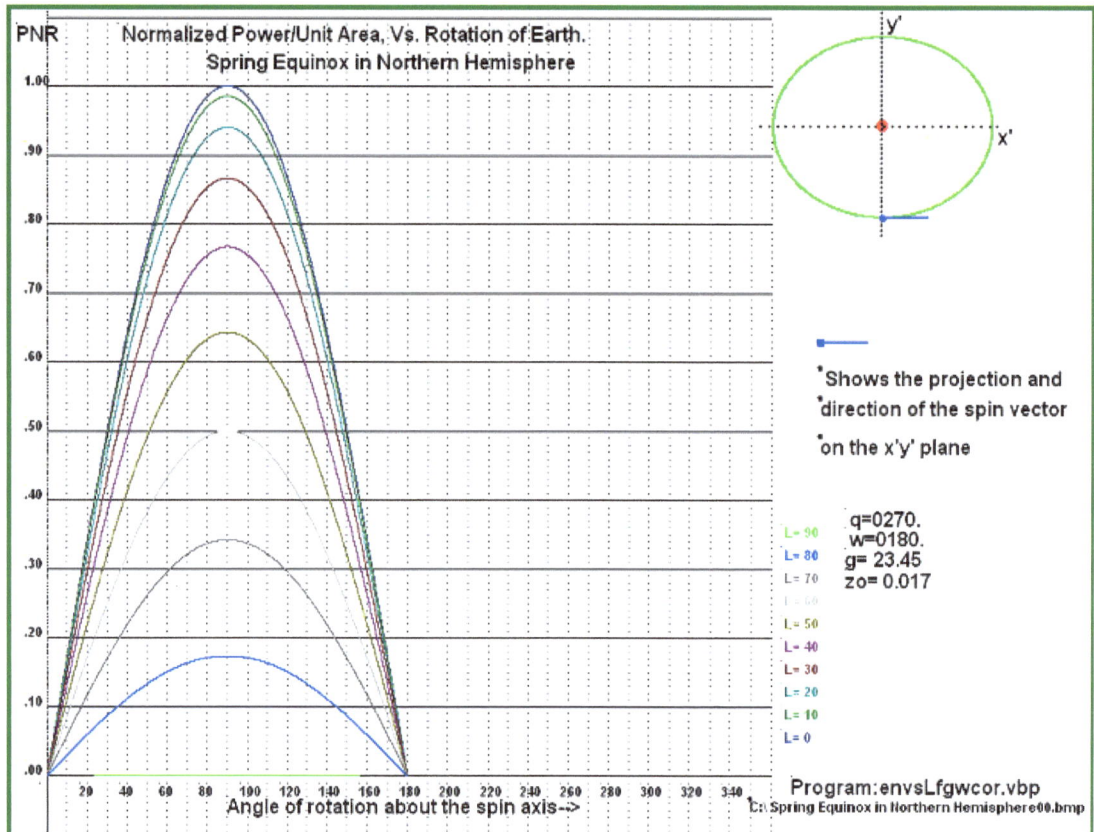

Figure (17)

Figure (17) shows PNR vs. rotation of Earth about the spin axis at Spring Equinox. The length of a day and night in both the Northern and Southern hemisphere is exactly 12 hours for all latitudes because the Sun's rays are normal to the spin axis. The North Pole (L=90) receives no power. The power received at high noon at the Equator is exactly equal to that received at Autumn Equinox.

Comparing the peak power received at the Equator in the winter to that received at the Equator in the summer we see from figures (15) and (5) that the peak power in the winter at the Equator is greater than the peak power at the Equator in the summer. This counter intuitive result is however predicted by equations (11) and (12).

To check the accuracy of equations (11) and (12), consider the diagram below which shows Earth's position at summer and in at winter Solstice.

22

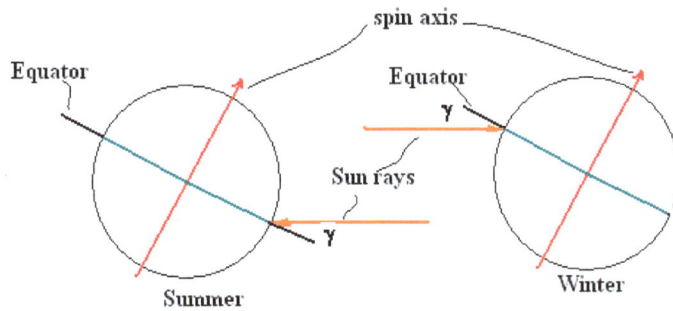

The normal component of the Sun's rays at the Equator in summer or winter is $\cos(\gamma)$. The normalized distance from the Sun is winter is $(1-z)$ and in the summer is $(1+z)$, hence, the

$$\frac{\text{maximum power received winter}}{\text{maximum power received summer}} = [(1+z)/(1-z)]^2$$

The geometry allows a check of the primary equations (11) and (12). Using the figure below and the current value of $\varepsilon=0.0170$ we find that if the latitude is zero, that is at the Equator, the ratio of maximum power received at winter to summer 1.070. From figures (5) and (17) we obtain[32] 1.070 as predicted by equations (11) and (12).

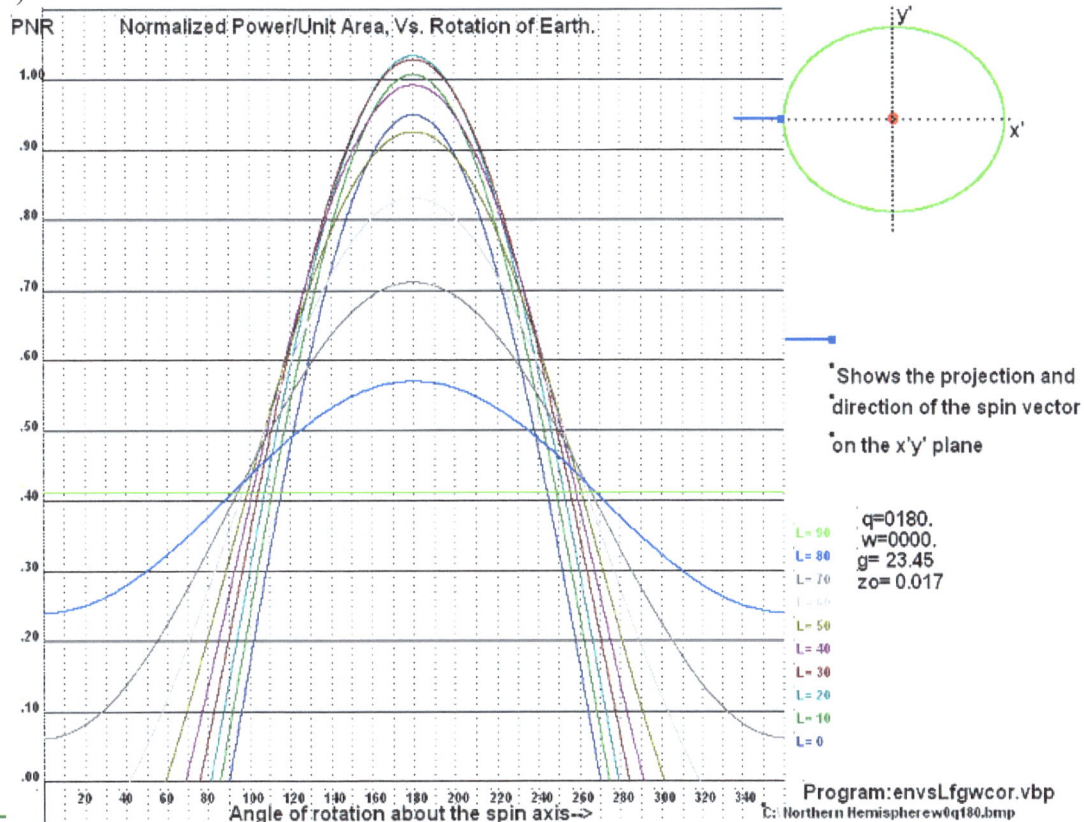

Figure (18)

Figure (18) shows PNR versus rotation of Earth about the spin axis when Earth is nearest to the Sun with the precession angle at 0^0 relative to the x' axis. Since the spin vector points toward the Sun and is nearest the Sun the power per unit area is the largest possible value. Under this configuration the season is 'Summer Solstice'. This 'summer' season receives more power from the Sun than the current configuration shown in Figure (5)

---

[32] Rounded to allowed number of significant figures.

23

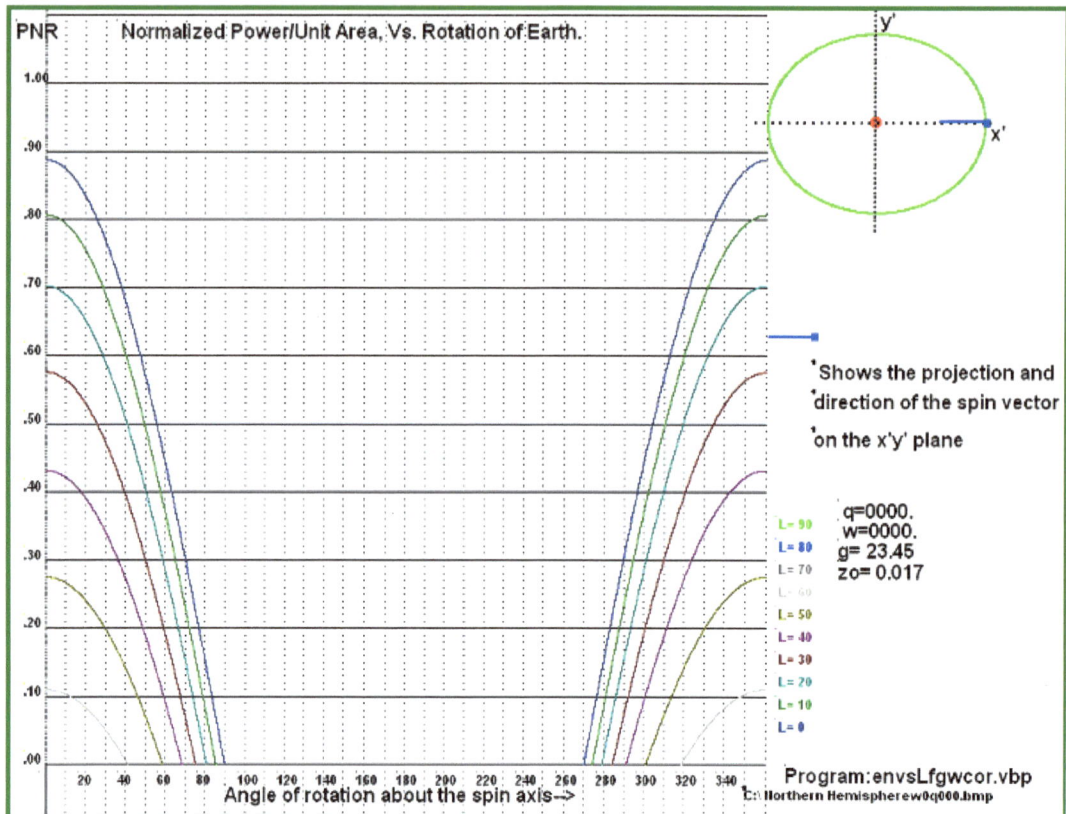

Figure (19),above, shows the winter season if w=0. The ratio between maximum power in the summer (See figure (18)) to power on the Equator in winter to maximum power in the summer is 1.070 as expected.

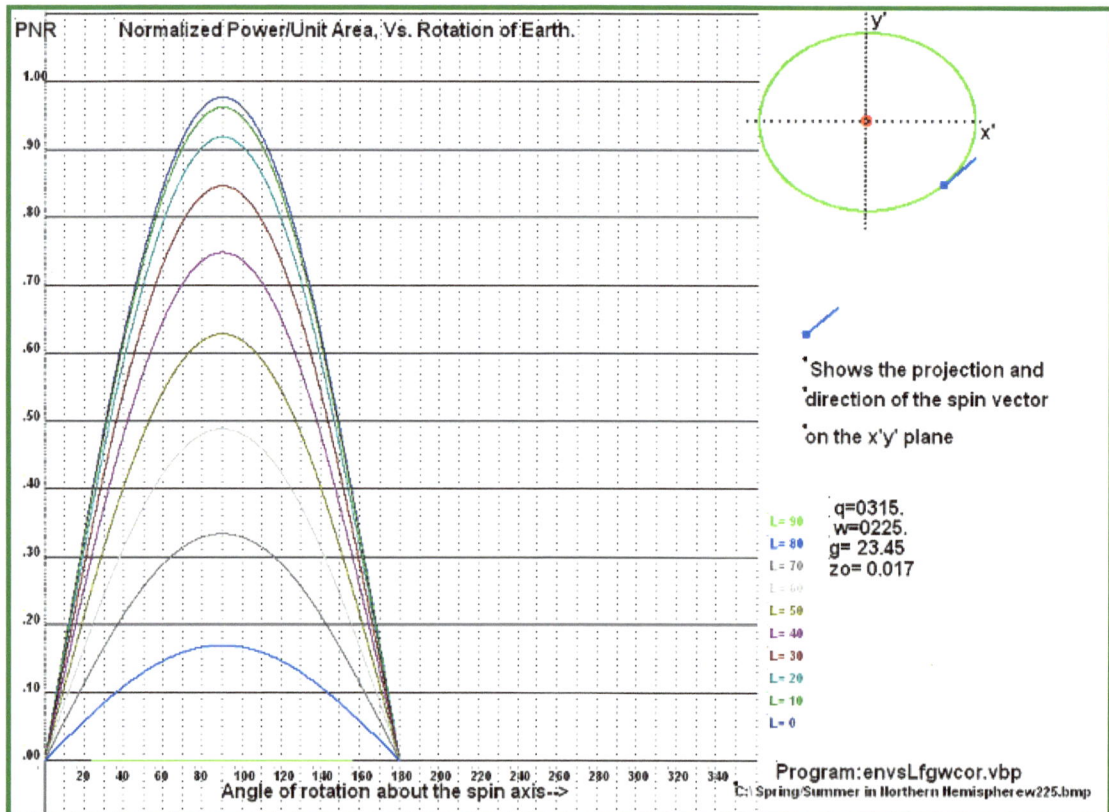

Figure (20), above, shows PNR vs. rotation of Earth about the spin axis when Earth is between Spring Equinox and Summer Solstice. The precession angle has rotated CCW 225°. The radius vector that

locates Earth is normal to the precession vector. At this particular configuration the length of a day for all latitudes is 12 hours.

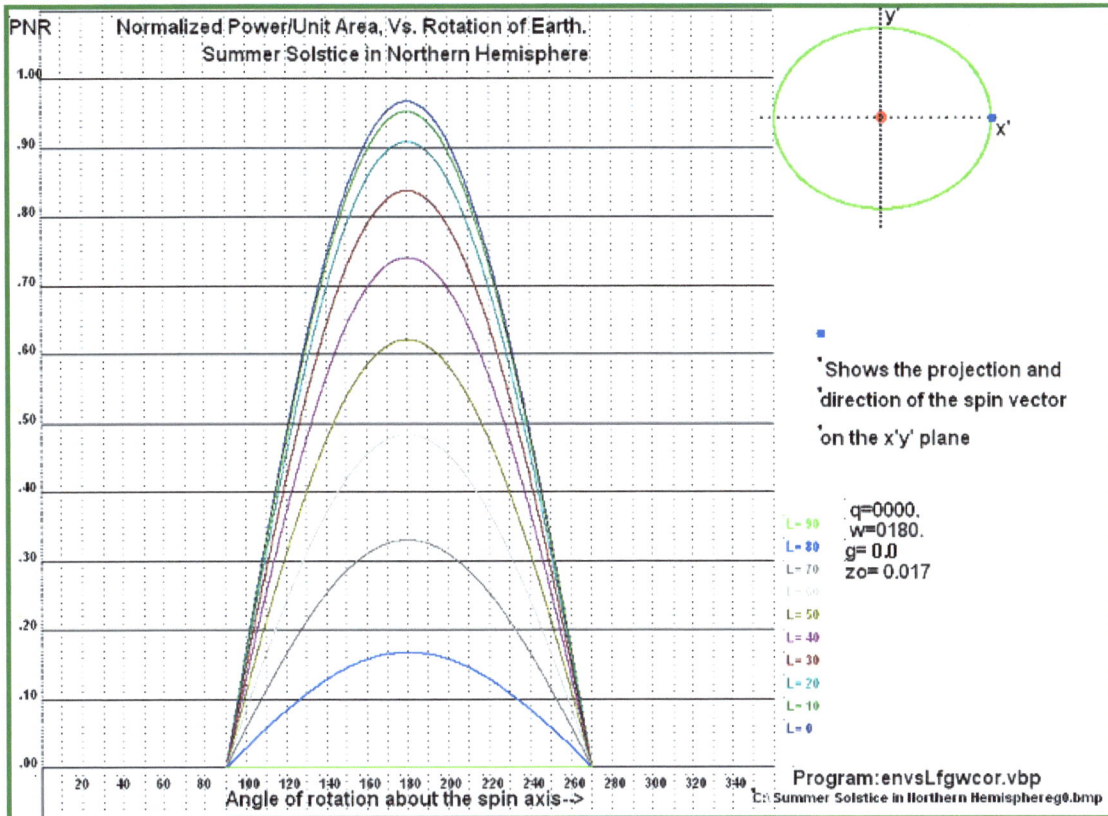

Figure (21)

Figure (21) shows Summer Solstice in the Northern Hemisphere when γ=0. The length of a day or a night is independent of the latitude and equal to 12 hours. There are no seasons; the energy received at a given latitude is always unchanged day to day, year after year.

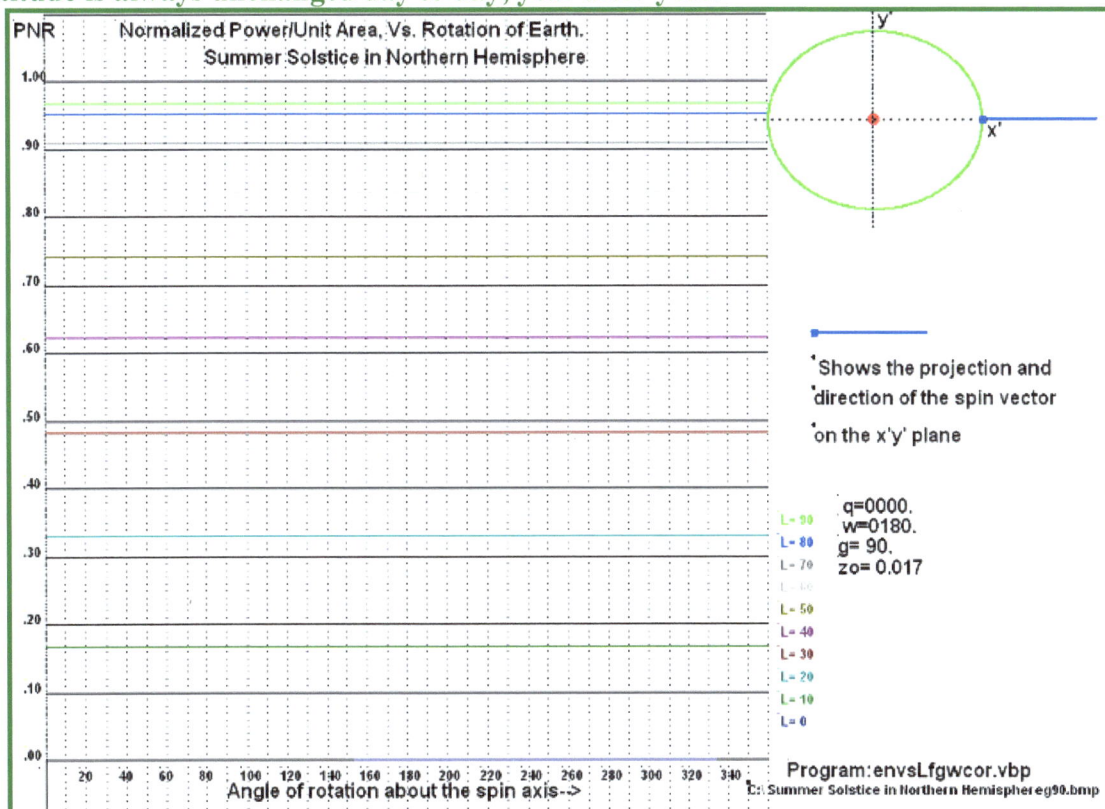

Figure (22) (See page 25) shows power received versus rotation of Earth about the spin axis if the spin axis of Earth is parallel to the orbital plane (γ=90). At a given latitude there is no variation in power received during the day and if θ=0 and ω=180 a day is 24 hours. The energy received increases as one moves from the Equator where <u>no</u> energy is received to a maximum energy received at the North Pole.

Figures(16) through (22) show that changes in the spin angle, γ, and the precession angle, ω, cause significant changes in energy received by Earth. Thus it is clear from the theory and satellite observations that the prevailing wind patterns in the Atlantic and Pacific oceans and the trajectories of hurricanes are caused by the distribution of energy received from the Sun. The energy received depends in turn on the inclination of Earth's spin axis relative to the orbital plane and the direction of the spin axis.

Since the principle source of energy received by Earth is from the Sun it is clear that understanding the effect of changes is Earth's orientation as it orbits the Sun is essential to an understanding of global temperature changes on Earth.

Chapter II

Effect of Precession, Tilt, and, Eccentricity on Power Received in a Year.

Earth rotates about a line connecting the North and South Poles called the spin axis. Because of rotation about the spin axis centrifugal forces develop that cause Earth to bulge in the vicinity of the equatorial plane. The resulting shape is that of an oblate spheroid. Because of this bulge and the tilt of the equatorial plane relative to the orbital plane the Moon and the Sun apply a torque to Earth. That torque causes counterclockwise precession of Earth's spin axis about a line normal to the orbital plane. Consequently, the time of an Equinox advances a small amount each year. In approximately 22,000 years[33] the spin axis will complete a 360° rotation about a normal to the orbital plane thus sweeping out a cone centered about a line always normal to the orbital plane. The angle between the **projection** of the spin vector onto the orbital plane and the sun system x' coordinate, ω, is used in this analysis to locate the position of the spin vector. The angle, ω, is called the precession angle. (See figure (2), Chapter (I).

Recently, Berger and Loutre have calculated[34] the eccentricity, ε, and the spin axis angle, γ, as a function of time (See figure (23) below).

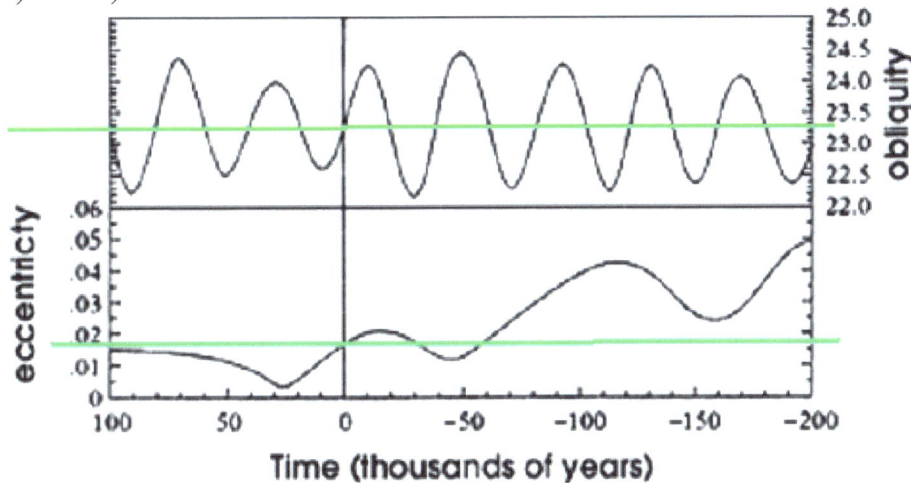

Figure (23)

Figure (23) shows the eccentricity, ε, and tilt angle, γ (obliquity), as a function of time. Zero on the time line corresponds to today and to the current precession angle, $\omega=180^{\circ}$, and, eccentricity. From A. Imbrie and K.P. Imbrie, but modified for this paper (See the footnote (35).

Since this monograph only applies to the last 22,000 years, only the most recent changes in eccentricity and tilt angle as calculated by Berger and Loutre are used.

Figure (24a) and (24b) show the portions of the tilt angle curve used. Since equations (11) and (12) describe the direction of the spin axis in terms of the precession angle, the data in figures (23) is re-plotted in terms of precession angle instead of time.

---

[33] The precession of the spin axis consists of two rotations. One counterclockwise rotation caused by the torque applied by the Sun and Moon and the second by a clockwise rotation of the Earth/Sun system about the center of mass of the system caused by gravitational interactions with the larger planets.

[34] A. Berger and M. F. Loutre Quaternary Science Reviews, Volume 10, Issue 4, 1991 pages 297-317
also J.Imbrie and K.P. Imbrie. Ice Ages , Solving the Mystery 1979, page 170

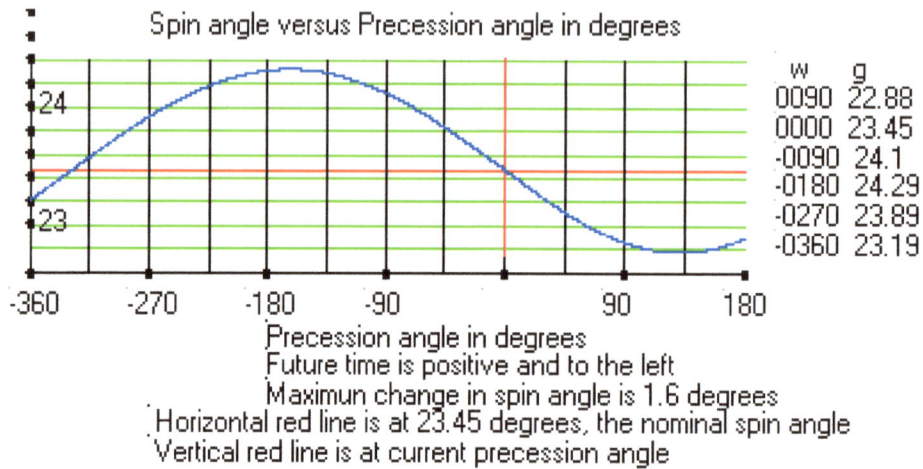

Figure(24a)

Figure (24) shows the tilt angle, γ, versus precession angle, ω. A precession angle of -360° corresponds to 22,000 years BP. A positive precession angle corresponds to a future time. Currently, the tilt angle, γ, is 23.45 degrees. To be consistent with the definition of the precession angle as illustrated in figure (2), add 180 degrees to the precession angles shown in figure (24a). The range of the variable, ω, is from -180 to +180 degrees. Referring to figure (27), this range is equivalent to the spin vector progressing from configuration F to G then to H,B,C,D, and returning to F.

The equations that fit the segments of the Berger and Loutre curves used are,

$$\text{If } \omega < \pi , \gamma = 23.45 - 0.85 * Sin(22 / 40 * (\omega - 0)) \text{ and}$$

$$\text{if } \omega > \pi , \gamma = 23.45 - 0.7 * Sin(22 / 32 * (\omega - 0)).^{35}............eqs.(13)$$

Corrected to account for the conventions used in figure (1) the equations become,

$$\text{If } \omega < \pi , \gamma = 23.45 - 0.85 * Sin(22 / 40 * (\omega - \pi)) \text{ and}$$

$$\text{if } \omega > \pi , \gamma = 23.45 - 0.7 * Sin(22 / 32 * (\omega - \pi))............eqs.(14)$$

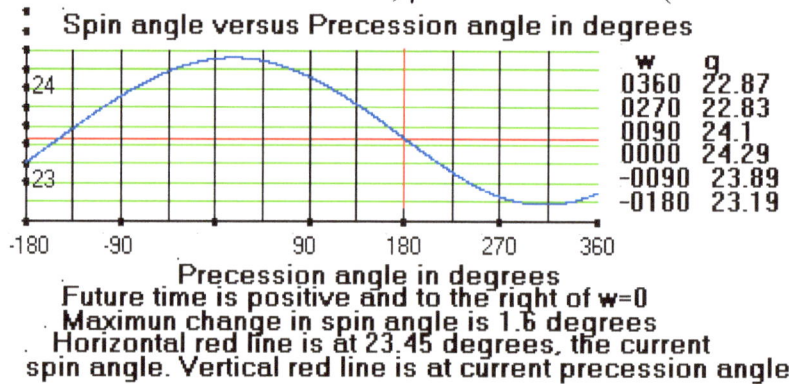

Figure (24(b))

A plot of the tilt angle, γ, versus precession angle, ω, after corrected for convention in figure (2).

At ω=0 the spin vector points to the left and is parallel to the x' coordinate. At ω=180 the spin vector points toward the right and is parallel to the x' coordinate. Calculated using eqs.(14).

Timex10⁻³ years.	ω	γ from Berger and Loutre	γ from eq(14)
11	180+180=360	22.8	22.87
7.5	90+45+180=315	22.7	22.8
5.5	90+180=270	22.8	22.83
2.5	45+180=225	23.05	23.1

---

[35] The equations give the tilt angle as a function of the precession angle instead of time directly as given by Berger and Loutre. The precession angle is a parametric function of time.

28

+0	180=180	23.45	23.45
-0		23.45	23.45
-2.5	-45+180=135	23.7	23.8
-5.5	-90+180=90	24.1	24.1
-7.5	-135+180=45	24.25	24.3
-11	-180+180=0	24.38	24.29
-16.5	-270+180=-90	23.9	23.89
-18.5	-315+180=-135	23.6	23.3
-22	-360+180=-180	23.08	23.19

Table (1)

Column 3 shows γ (from Berger and Loutre).Columns 1 and 2 show the time and corresponding precession angle as defined in figure (2). Column 4 shows γ calculated using eq. (14).

Eccentricity versus Precession Angle

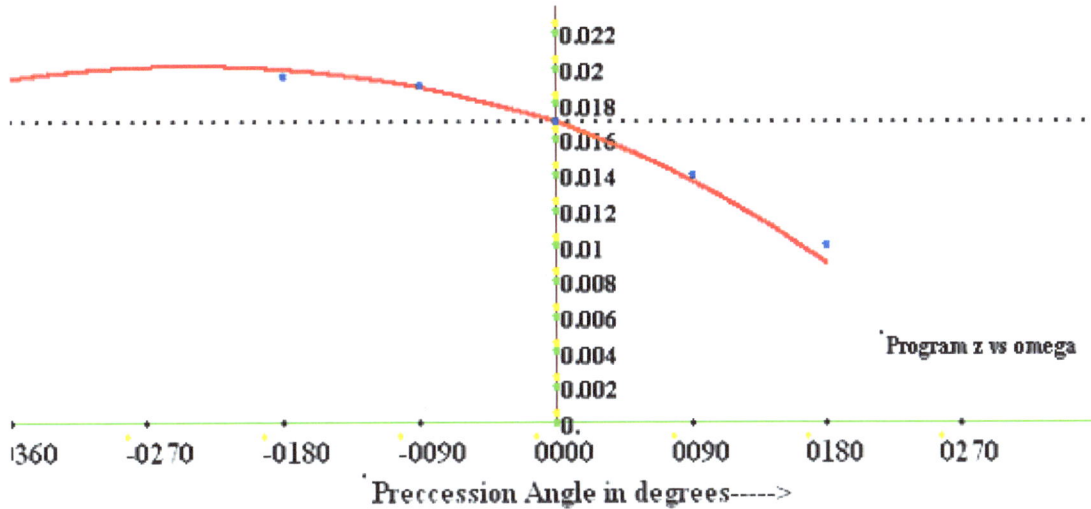

Figure (25)

The red curve is eccentricity versus precession angle over the period from approximately 22,000 years BP to 11,000 years after present calculated using equations (15). The curve is a best fit to the work of Berger and Imbrie. The eccentricity changes from 0.01 to 0.02 units in 22,000 years. The blue points are from Berger and Imbrie.

Corrected for convention in figure (2) the equations become;
if ω>π ,

$$\varepsilon = 0.017 - 0.004 / \pi * (\omega - \pi) - (0.003 / (2 * \pi^2)) * (\omega - \pi)^2,$$

and if ω<π ,

$$\varepsilon = 0.017 - 0.004 / \pi * (\omega - \pi) - (0.0055 / (4 * \pi^2)) * (\omega - \pi)^2$$

.............eqs.(15)

Variables Determining the Total Energy Received by Earth in a Year

From figure (24b) we see that in the near future the energy received at the high latitudes decreases as the tilt angle decreases suggesting a cooling trend going forward in time. However, the total energy received depends not only on the magnitude of the tilt angle but also on the direction of the tilt angle and eccentricity of Earth's orbit. Thus to determine the total energy received one must calculate the total energy received resulting from the combined effects of changes in tilt and precession angles, and eccentricity.

Equations (11) and (12) in Chapter I describe the power received by Earth as a function of, ϕ, L, ω, γ, θ, and, ε. The longitude, ϕ, is used as the independent variable to show how the power received changes in a day. To obtain the energy received in a day we must sum the power/unit area times the incremental area, ΔA on Earth, times the corresponding increment in time, Δt, that that increment of area receives power from the Sun. To obtain the energy received in a year we must advance the position of Earth in orbit by a small increment, $\Delta\theta$, and repeat the sum described above. After repeating the steps above and summing the energy received after each step until Earth has completed an orbit about the Sun we will have obtained the total energy received for that particular year. Then, ω, is increased by an increment, $\Delta\omega$, and all the steps above are repeated to obtain the energy obtained for the next value of ω. Thus time used in this analysis is measured in terms of a unit of angular displacement of the projection of the spin axis onto the orbital plane.

Since Earth's precession angle completes a 360° rotation in 22,000 years, units of time are directly related to the precession angle, ω. Thus a unit of time, Δtime, equals $22,000)/(2\pi)*\Delta\omega$ years.
Positive ω is the CCW angular displacement of the projection of the spin axis on the orbital plane.

Earth's Orbit and Measurement of Time.
Because Earth is acted upon by a force directed along a line connecting the center of mass of Earth to the center of mass of the Sun, the angular momentum vector of Earth about the Sun is constant in magnitude and direction and always normal to the orbital plane. Given the above, the area circumscribed by Earth's orbit year after year, is also constant provided the eccentricity remains constant.

The angular momentum of Earth about the Sun is $\mathbf{M}=m\mathbf{R}\mathbf{x}\mathbf{v}$ where m is Earth's mass and $\mathbf{v}$ is its orbital velocity. The torque,τ, is $d/dt(\mathbf{M})=m\mathbf{R}x(d^2\mathbf{R}/dt^2)+m\ (d\mathbf{R}/dt)x(d\mathbf{R}/dt)=0$, thus $\mathbf{M}=\mathbf{constant}$[36]. Since $\mathbf{M}dt=m\mathbf{R}x dx=md\mathbf{A}=\mathbf{constant}*(dt)$, the area swept out is a linear function of time or stated differently the time to complete an orbit is a linear function of the area swept out by the radius vector.

If the major axis remains constant as the eccentricity changes, the area circumscribed by Earth's orbit changes, as does the time to complete an orbit. Thus, time used in this analysis is based on the time for the spin axis to precess 360° and _not_ the time for Earth to complete a single orbit. In 22,000 years, the eccentricity decreases from 0.02 to 0.01 causing the time to complete one orbit (a year) to increase by ~0.04%.[37]

 Figure (26) shows the area swept out by the radius vector as it swings from $\theta=-180^\circ$ to 180°. Since areas are swept out are a linear function of time it is clear that it takes Earth longer to complete the -180° to 0° sector of the orbit (the black curve) than more eccentric (blue and green) orbits. Hence, if the eccentricity changes over time, the conventional definition of a year as the time for Earth to complete one orbit about the Sun should is not used as a unit of measure of time.
Even though the total time to complete an orbit over the time period studied in this analysis is very small change, the unit of time used in this analysis is not the current year. Instead, the unit of time is reckoned in terms of, ω, the angular displacement of the spin axis vector about a vector normal to the orbital plane (i.e. the precession angle). However an assumption is made that the orbital plane remains unchanged in direction.[38]

---

[36] Remember Force=$m(d^2\mathbf{R}/dt^2)$ and is along a line connecting Earth and Sun centers.
[37] The area of the ellipse is equal to $\pi a^2(1-\varepsilon^2)$, where ,a, is the length of the major axis and, ε, is the eccentricity. In this monograph ε changes from 0.02 to 0.017 unit or and area change of ~0.04%.
[38] Earth's angular momentum, M, equals $mR^2\omega_e$ =constant, and, m= the mass of Earth. Since Earth's orbit is eccentric R increases or decreases as Earth orbits the Sun causing the angular velocity to decrease or increase, respectively, to maintain the angular momentum constant..

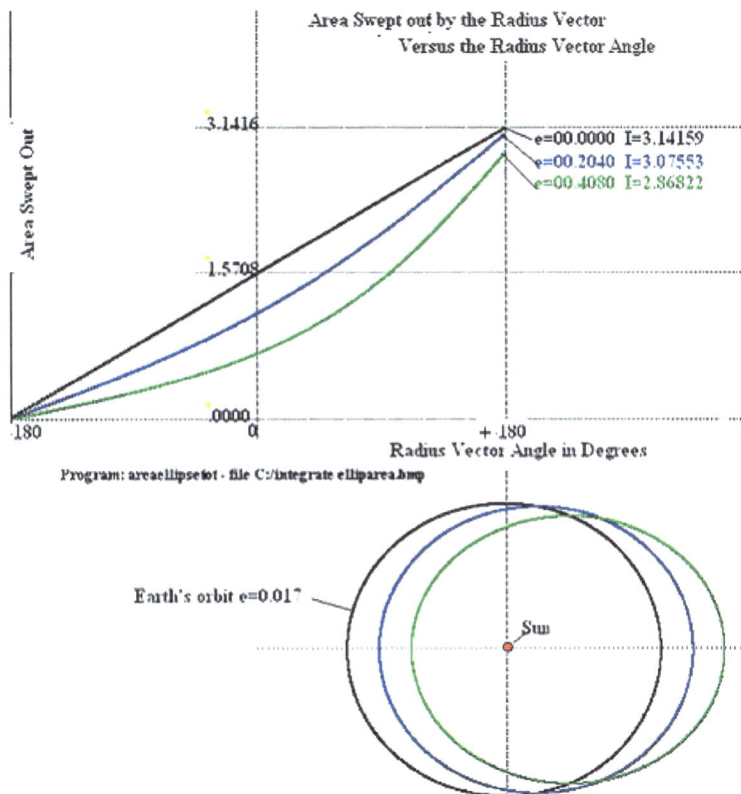

Figure (26)
Shows the area, I, swept out by the elliptical orbit radius vector, $R=a(1-\varepsilon^2)/(1-\varepsilon Cos(\theta))$, where, a=major axis, ε= eccentricity, and, θ radius vector angle. The eccentricity changes from 0.000 to 0.4080.

Using equations (11) , (12), and, the dependence of eccentricity and spin axis angle on precession angle, we can obtain the power or **the total *energy received per given year or sequence of years*** by Earth in the past or in the future. In addition, we can determine the energy received in only the current summer or winter months or the energy received as Earth travels through any sector of its orbit about the Sun. Since equations (11) and (12) predict the power received as a function of the angle of precession, ω, we can calculate the energy received some 2000 years ago or predict the energy received any arbitrarily chosen time forward and for any configuration of precession angle, tilt angle, eccentricity, and, latitude[40].

Figure (27) shows, schematically, the spin axis direction[41] as the spin axis vector precesses counterclockwise about a line normal to the orbital plane. According to the coordinate system shown in figure (2), the configuration if $\omega=0$ or $360°$ is depicted in figure (27C). As Earth travels around the Sun in a given unit year, the direction and magnitude of the spin vector remains essentially unchanged because the spin angular momentum of Earth is essentially constant during a given year. If $\omega=0$, Equinox occurs if, $\theta=90°$, or, $270°$, respectively, and the energy received at Autumn Equinox is the same as that received at Spring Equinox. Summer and Winter Solstice occur if $\theta=180°$ and, $\theta=0°$, respectively. At Summer Solstice the power received by Earth is greater than for today's configuration which is shown in figure (27F). As ω continues in a counterclockwise direction the direction of the spin axis changes to the configuration shown in figure (27D) and then, to the current configuration shown in figure (27F), the current configuration. After going through

---

[40] Using Berger and Loutre one could calculate the energy back some 600,000 years, but the purpose of this work is to predict the recent (the last 22,000 years) and future (5,000 years) trend in energy received.
[41] The projection of the spin axis onto the orbital plane is shown. The magnitude of the spin axis vector is held constant over the period of one year.

configurations shown in Figures (27E) and (27F) today's configuration will begin again. It is clear that as Earth progresses through the various configurations the energy received by Earth differs from Solstice to Solstice and Equinox to Equinox.

Since the spin axis completes a single precession in 22,000 years it will have completed, in the last 2009 years, a $(-2009/22,000)*360 = -32.9$ angle of precession. Earth's spin axis direction will have followed a counterclockwise circular path on the celestial sphere from $\omega=180°- 32.9°=147.1°$ to the current location and the tilt angle changed from $23.85°$ to the current angle of $23.45°$.

Referring to figure (2), Chapter I, we see that according to the convention adopted in figure (2) the precession angle, ω, was at zero degrees, $(-180/360)*22,000=-11,000$ years ago or 8991 B.C.. Referring to Figures (2) and (3) the precession angle, today, is approximately $+180°$ and the spin axis is tilted toward the Sun. When $\theta=0°$, and the Sun is directly overhead at $23.45°$ north latitude at Summer Solstice. In the year 3492BC (-5500 years ago), ω was $+270°$, Winter Solstice occurred at $\theta=270°$ and Summer Solstice occurred at $\theta=90°$. Autumn and Spring Equinox were at $\theta=0°$ and $\theta=180°$, respectively, and unlike today the power received at high noon at the Equinoxes was not equal.

Clearly, energy from the Sun received by Earth depends not only on Earth's position in orbit, but also on the angle of precession, the tilt of the spin axis, and, the eccentricity of the orbit. Figure (27) shows schematically, Earth's elliptical orbit and six orbital positions of Earth with the spin axis rotated counterclockwise from today's configuration.

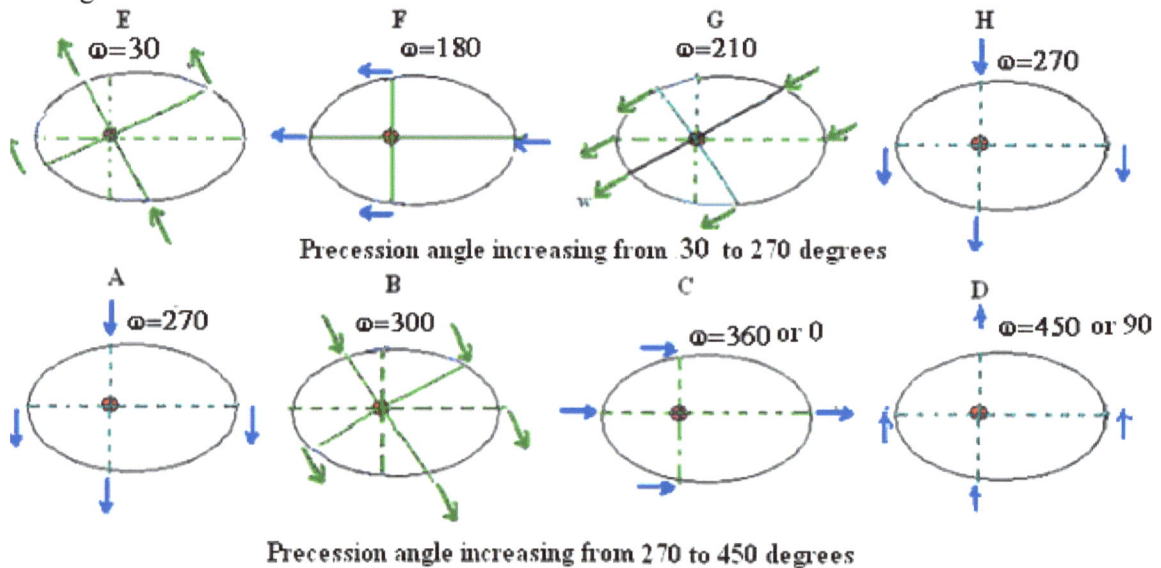

Precession angle increasing from 30 to 270 degrees

Precession angle increasing from 270 to 450 degrees

Figure (27)

Figure (27) shows schematically, Earth's elliptical orbit and six orbital positions of Earth with the *projection* of the spin axis on the orbital plane rotated counterclockwise from today's configuration. The displacement of Earth in orbit, θ, is measured relative to the positive x' Sun system coordinate[42]. Figure (27F) is the current configuration ($\omega=180°$). The spin axis is pointing toward the Sun at Summer Solstice and away from the Sun at Winter Solstice. In figure (27C) at Summer Solstice, Earth is closer to the Sun than at the prior Summer Solstice shown in figure (27F), and summer is therefore warmer than for the configuration in figure (27F). At Winter Solstice it is colder than in the figure (27F) configuration because Earth is further from the Sun. In figure (27A) Autumn Equinox is warmer than Spring Equinox.

One might conclude that the average energy received for the year shown in figure (26C) is larger than the configuration of figure (27F). Since both, γ and ε are functions of ω, to be certain one must calculate precisely the total energy received in the years depicted in the figure (27C) and (27F) configurations.

---

[42] See figure (2), Chapter I.

Energy Received from the Sun as a Function of Precession, Latitude, and, Spin Axis Tilt Angles, and, Eccentricity.

Equations (12) give the normalized power per unit area received by Earth as a function of Earth's spin axis tilt angle, precession angle, latitude and longitude, and the position of and distance of Earth from the Sun. Note that the quantity, P_{NR}, is the power per unit area normal to Earth at a specified latitude, precession angle, etc. Let, ΔA, be a small surface area of Earth located at the specified latitude and longitude that is receiving the incoming power. That area is given by the expression,

$$\Delta A = R_E^2 \, Cos(L) \, \Delta L \Delta \phi = R_E^2 Cos(L) \, \Delta L \Delta \phi .$$

.................eq.(15)

R_E, is the radius of Earth including the atmosphere, L, is the latitude in the Northern Hemisphere, ΔL, a small angular displacement of L, and, $\Delta \phi$, is a small displacement in the longitude.

The time lapsed as Earth rotates $\Delta \phi$ radians is $\Delta \phi / \omega_{spin}$ where ω_{spin} is the angular velocity of Earth about the spin axis in terms of radians per year and $\Delta \phi / \omega_{spin}$ is the time in years for Earth to revolve $\Delta \phi$ radians. Hence, the product of the power/unit area and $\Delta A(\Delta \phi / \omega_{spin})$ is the *energy transmitted* from space into the small segment of area, ΔA, during a time Δt. Using equation (12) we write,

$$P_{NR} * \Delta A / \omega_{spin} = (1 - \varepsilon * Cos(\theta))^2 / ((1 - \varepsilon^2)^2 * P_N * R_E^2 * Cos(L) * \Delta L * \Delta \phi / \omega_{spin}$$

........eq.(16)

Total Energy Received in One Day by a Band of Earth's Surface.

The total energy received by the band that is $R_E \Delta L$ wide and wraps 360 degrees around Earth at latitude, L, at a specific orbital position, θ, is,

$$E_{band}(\varepsilon, \gamma, \theta, \omega, L) = \frac{\Delta L}{\omega_{spin}} \int P_{NR}(\varepsilon, \gamma, \theta, \omega, L, \phi) R_E^2 Cos(L) \, d\phi$$

....eq (17)

The limits of integration are from $\phi = 0$ to 2π. The width of the band is arbitrarily chosen as 10 miles, hence ΔL is 0.0025 radians (0.143 degrees).

Thus for specific values of, ε, γ, θ, **L**, and, ω, one can calculate the *total energy received in one day over a small band of Earth's surface centered about a specific latitude and at a specific orbital position of Earth* [43]. The integral is approximated by the sum,

$$E_{band}(\varepsilon, \gamma, \theta, \omega, L) = (\Delta L / \omega_{spin}) \Sigma P_{NR}(\varepsilon, \gamma, \omega, \theta, L, \phi) \, R_E^2 Cos(L) \, \Delta \phi . \quad ....eq. (18)$$

The increment chosen in longitude, $\Delta \phi$, is $2\pi / 360$ radians ($1°$) and the limits of the sum are from $\phi = 0$ to $\phi = 2\pi$.

The program used to calculate the energy received for a particular angle of precession follows. The equations, according to Berger[44], that relate the tilt angle and eccentricity to precession angle over the range of 22,000 years [45] are included in the software. To simplify programming we normalize the energy received in a band by setting[46] $R_E = 1$. In addition, energies calculated are normalized to the energy received at the current Summer Solstice when $\omega = 180°$ and at $L = 65°$ North latitude.

---

[43] This is the total energy flowing toward Earth at the beginning of the atmosphere.
[44] see page 28
[45] The rate of precession is not strictly constant, and has been reported between 22,000 and 23,000 years.
[46] Remember R_E is the radius of Earth and is not to be confused with R the distance between Earth and the Sun.

Software to Compute Total Energy Received per Year Versus Precession Angle.
The software described In Appendix I is used to compute the total energy received by Earth as it orbits from 270° to 90° and as it orbits from 90° to 270° in a year as a function of latitude, L, and the precession angle, ω.

The software computes E_{band} ($\epsilon,\gamma,\theta, \omega, L$) by[47],
(1) Evaluating $\Delta L \Sigma P_{NR}(\epsilon,\gamma,\omega,\theta,L, \phi)$ $\mathbf{R_E}^2\mathbf{Cos(L)}\Delta\phi$ for fixed value of $\epsilon,\gamma,\omega,\theta$, and, L as Earth steps in units of $2\pi/360$ radians or 1° about the spin axis. The position of Earth in orbit is located by θ (see figure (28)).

(2) The orbital position, θ , is advanced $2\pi/364$ radians[48] or 0.9890° and E_{band} ($\epsilon,\gamma,\theta, \omega, L$) is computed again. This result is added to the prior computation. Steps (1) and (2) are repeated until the sum is evaluated for the specified orbit of Earth about the Sun.

(3) The precession angle, ω , is then advanced by an increment of 5° and the sequence above repeated. Thus the energy received at each 5° increment of the precession angle is computed. Since ϵ and γ are known functions of the precession angle, the eccentricity and tilt angle change for each increment in ω.

The energy received is calculated for a specific trajectory and precession angle of Earth by adding the energy received each day to the prior day of the trajectory.
Thus, the energy received in a day is summed and the total energy received each day for all the days of the specified trajectory are added. As discussed before, summing over increments in ϕ is equivalent to summing over time because ϕ/ω_{spin} has units of time. The angular velocity of Earth about the spin axis is equal to $2\pi/[(24)(60)]=0.004363$ radians/minute. In the software the increment in ϕ is $2\pi/360$ radians, hence the energy received in a day is calculated at 4 minute intervals. The trajectories are defined in figure (28). The blue trajectory is between $\theta=90$ and 270 degrees. The red trajectory is between $\theta=270$ and 90 degrees. Thus the energy received in the red trajectory for a specific value of ω is the energy received as Earth travels from $\theta=270$ to 90 degrees. Similarly, the energy received in the blue trajectory for a specific value of ω is the energy received as Earth travels from $\theta=90$ to 270 degrees. For each value of ω, the orientation of Earth between $\theta=270$ to 90 degrees is not the same. If the precession angle changes, the total energy received for a year changes.

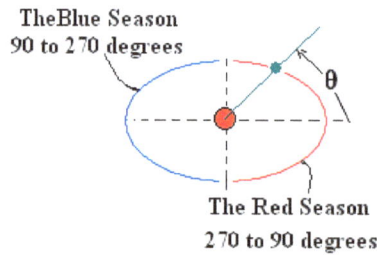

TheBlue Season
90 to 270 degrees

The Red Season
270 to 90 degrees

Figure (28)

Shows the definition of the trajectories used in the program to calculate energy received, E_{band} ($\epsilon, \gamma, \theta, \omega, L$),(see equation (17)) in a band ΔL wide centered about the latitude, L. At each increment in orbital position, θ, the energy received at each increment in ϕ is summed and added to the energy received at the prior orbital position. The process continues until the energy received at all orbital positions between $\theta=270^\circ$ and 90° for the given precession angle are summed. That sum is the total energy received as Earth moves from $\theta=270$ to 90 degrees. Similarly, energy received as Earth moves from $\theta=90^\circ$ to 270° is summed and that sum is the energy received as Earth moves from $\theta=90^\circ$ to 270°.
As shown the red season is the current summer season if $\omega=180^\circ$. If $\omega =180^\circ$ and $\theta=270^\circ$ to 90° then the blue season is a winter season.

---

[47] See equation (18)
[48] A 364 day year is used to simplify the calculations.

Energy Received by Earth per Year as a Function of the Angle of Precession and Latitude.

Figures (29) and (30) show, as calculated by the program, the normalized energy received by Earth versus the angle of precession and latitude. The red curves show the energy received as Earth travels from $\theta=270^\circ$ to 90° at a given latitude versus precession angle. The blue curves show the energy received at a given latitude versus precession angle as Earth travels from $\theta=90^\circ$ to 270°. The increment in precession angle is 5° which is equivalent to 305.6 years[50]. The current precession angle, 180°, is indicated by the vertical green line. The energy received[51] is shown for north latitudes, starting at L= 20° and ending at L= 80°. The red and blue numbers on the upper right are the numerical values of the energy received normalized to the value at ω = 180, and at L=65 degrees and are used later to calculate $\Delta E/\Delta t$ in equation (19). Since the program uses a 364 day year the time between $\omega=0$ and $\omega=180$ is shortened by approximately 56 years and the entire 22,000 year cycle by approximately 75 years. Thus one should add 0.8° to the precession angle (or 88 years) at $\omega=180^\circ$.

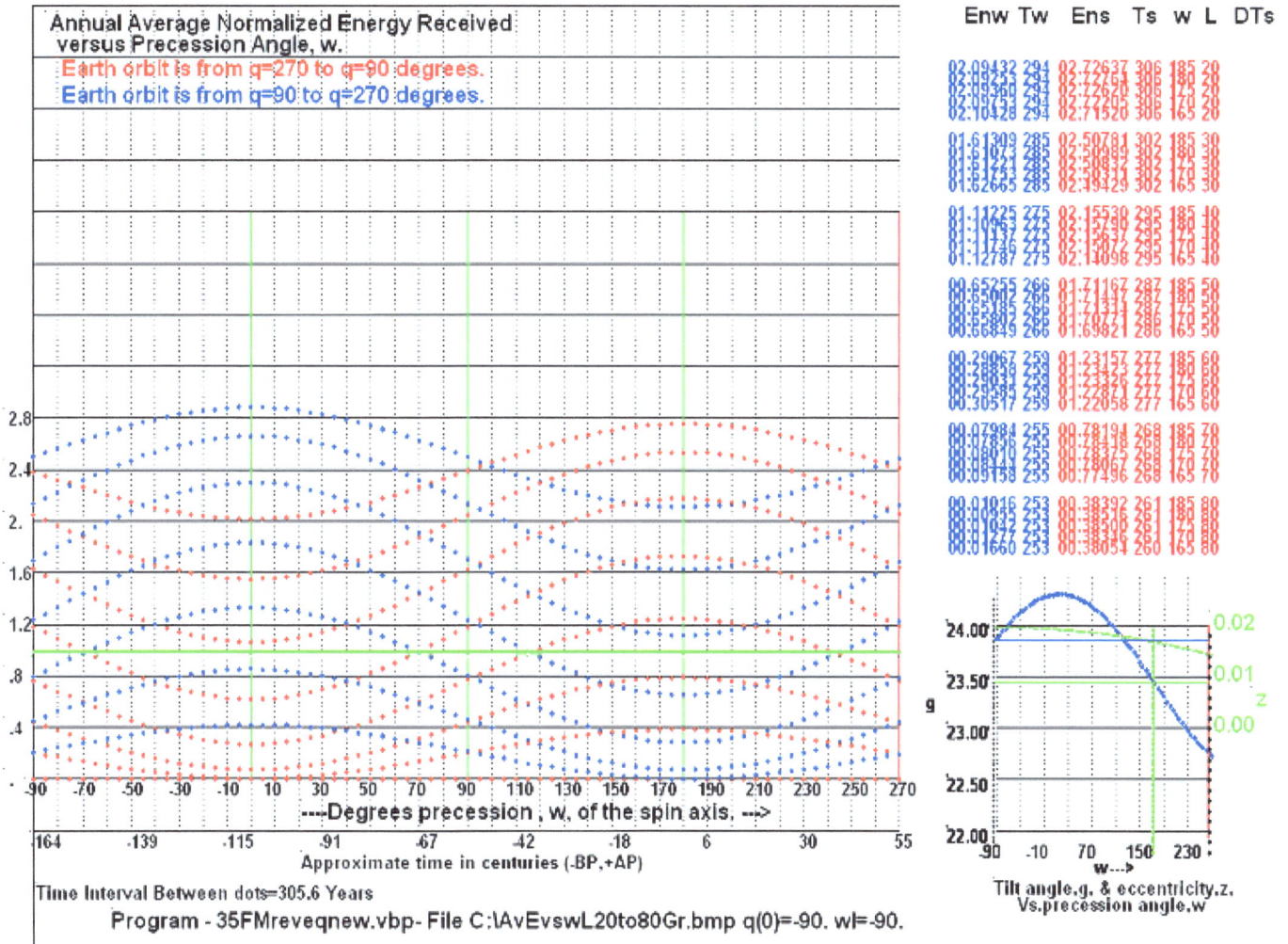

Figure (29)

---

[50] The time used for a complete cycle of the precession angle is 22,000 years a rotation of the projection of the precession angle from $\omega=0$ to $\omega=360^\circ$.

[51] Energy received is distributed over a narrow band centered about the indicated latitude.

Figure (29) shows on the vertical coordinate, the energy received from the Sun, in a band centered about a given latitude, normalized to the energy received at ω=180° and L=65° in the Summer[52]. The precession angle is plotted on the horizontal coordinate as is the approximate time BP and AP in centuries. The tilt angle, γ, and eccentricity, ε, are shown in the lower right versus the precession angle. The horizontal blue and green lines show the current values of, γ, and, ε, respectively. The red and blue numbers on the upper right are the numerical values the energy received normalized to the value at ω = 180°, and at L=65° degrees, and are blue as θ goes from 90° to 270° and red as θ goes from 270° to 90°. Note that at ω=0° the blue curves represent a summer season with maximum energy received at ω=0° and that the tilt angle is maximum.

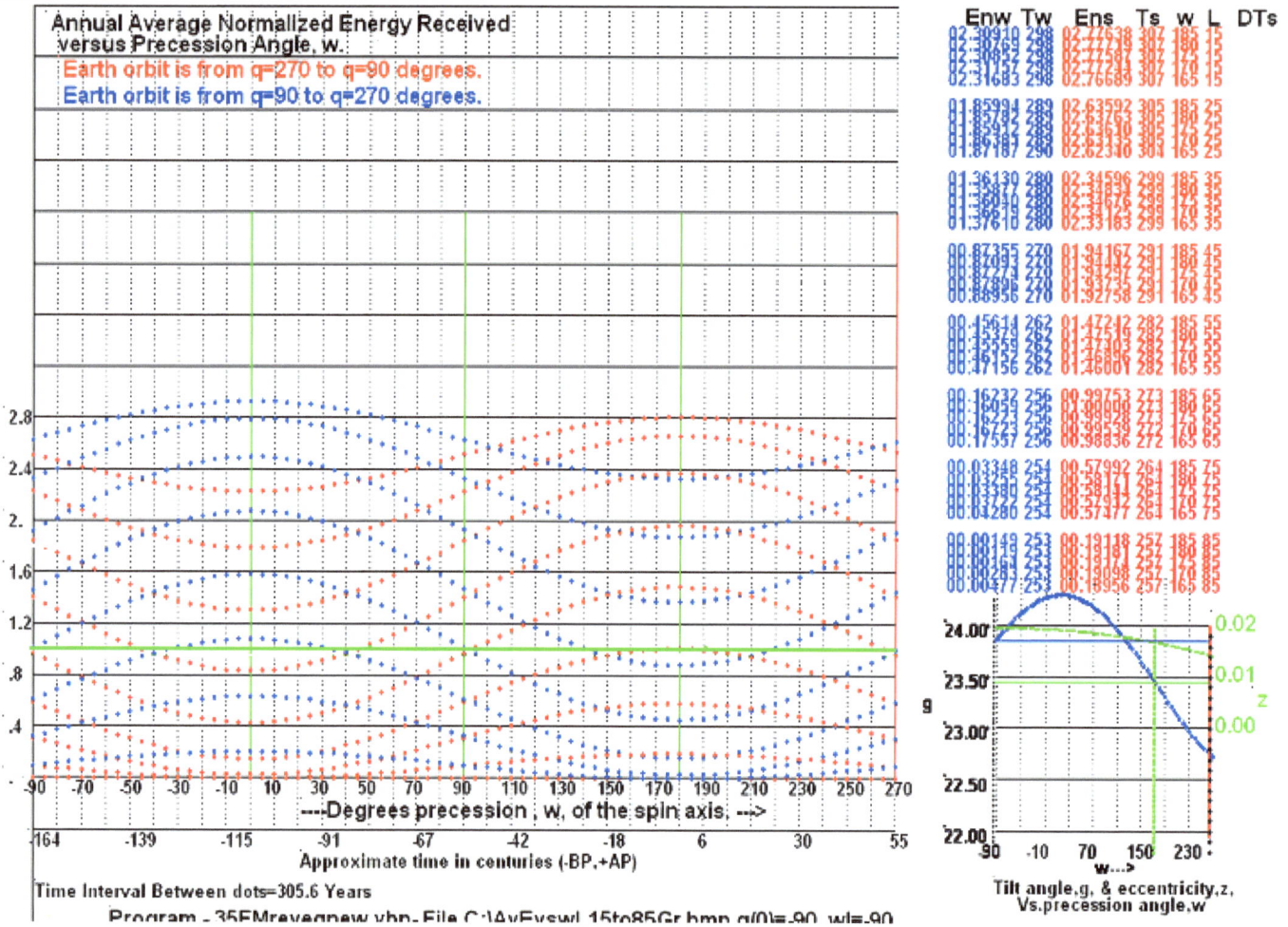

Annual Average Normalized Energy Received versus Precession Angle, w.
Earth orbit is from q=270 to q=90 degrees.
Earth orbit is from q=90 to q=270 degrees.

Enw Tw Ens Ts w L DTs

---Degrees precession , w, of the spin axis; -->

Approximate time in centuries (-BP,+AP)

Time Interval Between dots=305.6 Years

Program - 35EMrevegnew.vbp- File C:\AvEvswl 15to85Gr.bmp q(0)=-90 wI=-90

Tilt angle,g, & eccentricity,z, Vs.precession angle,w

Figure (30)

Shows the energy received versus precession angle for latitudes between 15 to 85 degrees. This data supplements the data shown in figure (29). Note that as the precession angle rotates beyond 180° and the spin axis tilt angle decreases the energy received passes through a maximum and declines. As shown later (See figures (38) and (39)) pages (47) and (49) this implies constant or declining surface temperatures at all northern latitudes in the future. Recent temperature measurements obtained from microwave sounding units on satellites (University of Alabama)[57] appear to support this prediction.

---

[52] Using energy received at ω=180° and L=65° compares all energy points to energy received currently near the Arctic Circle.

[57] S. Fred Singer, *Nature, Not Human Activity, Rule the Climate Report of the NIPCC*, Heartland Institute, pg.,10, Figure(13)

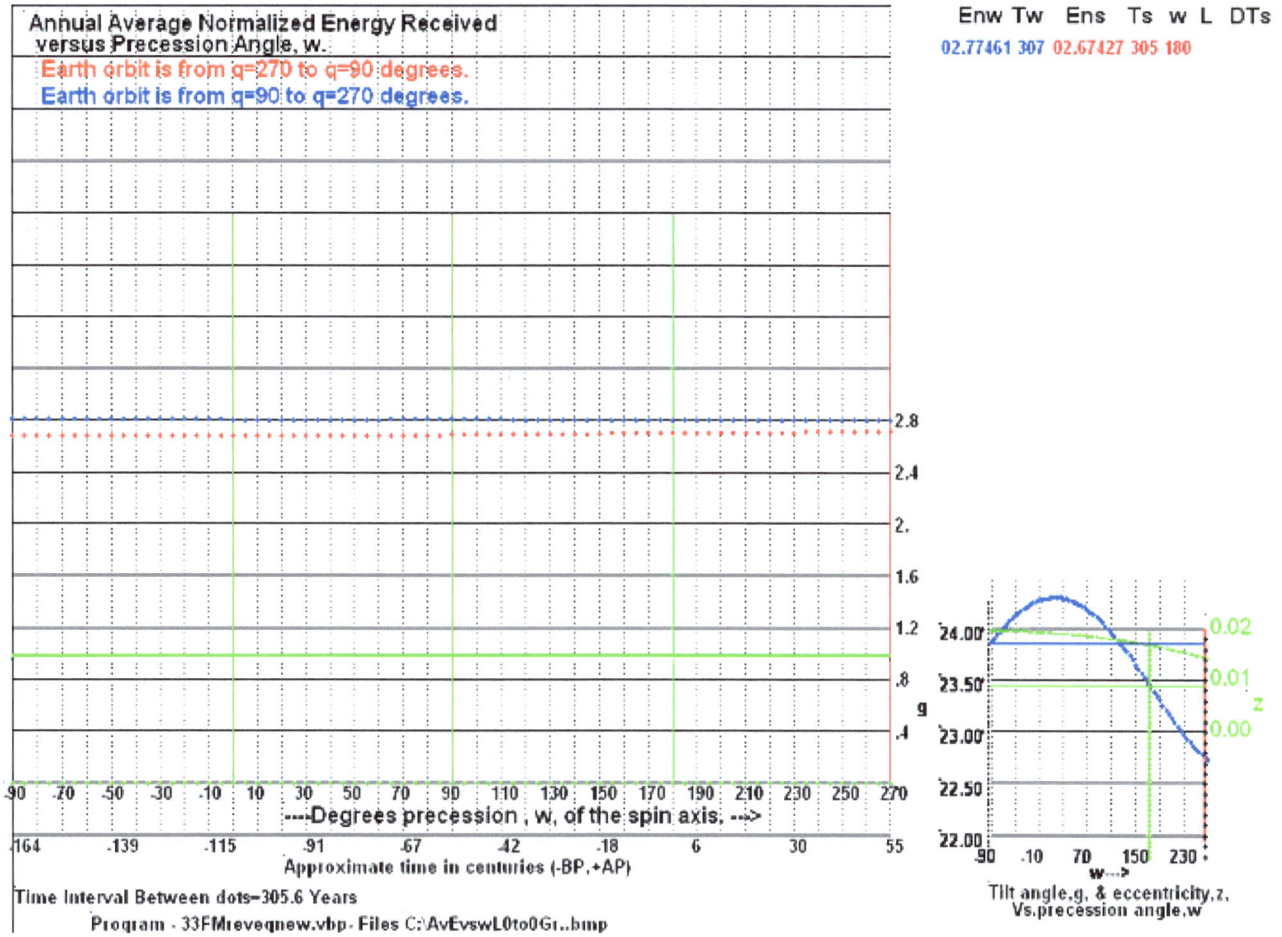

Figure (31)

Figure (31) shows the energy received in a ribbon centered about the Equator. The energy received is nearly independent of the precession angle and is greater if Earth's trajectory is between $\theta=90^0$ to 270^0 (the blue curve) than when the trajectory is between 270^0 and 90^0 (the red curve). The difference between the energy received between the blue and red trajectories is greatest when the spin vector points mostly away from the Sun and Earth is nearest the Sun. Note that if $\omega=0^0$ the tilt of the spin vector is greatest. If $\omega=180^0$ the tilt angle is less by 0.8 of a degree and the distance to the Sun is less accounting for a further reduction in Sun light intensity. Thus we see a competing and counter intuitive relationship between the variables controlling the energy received by Earth. (See page (23))

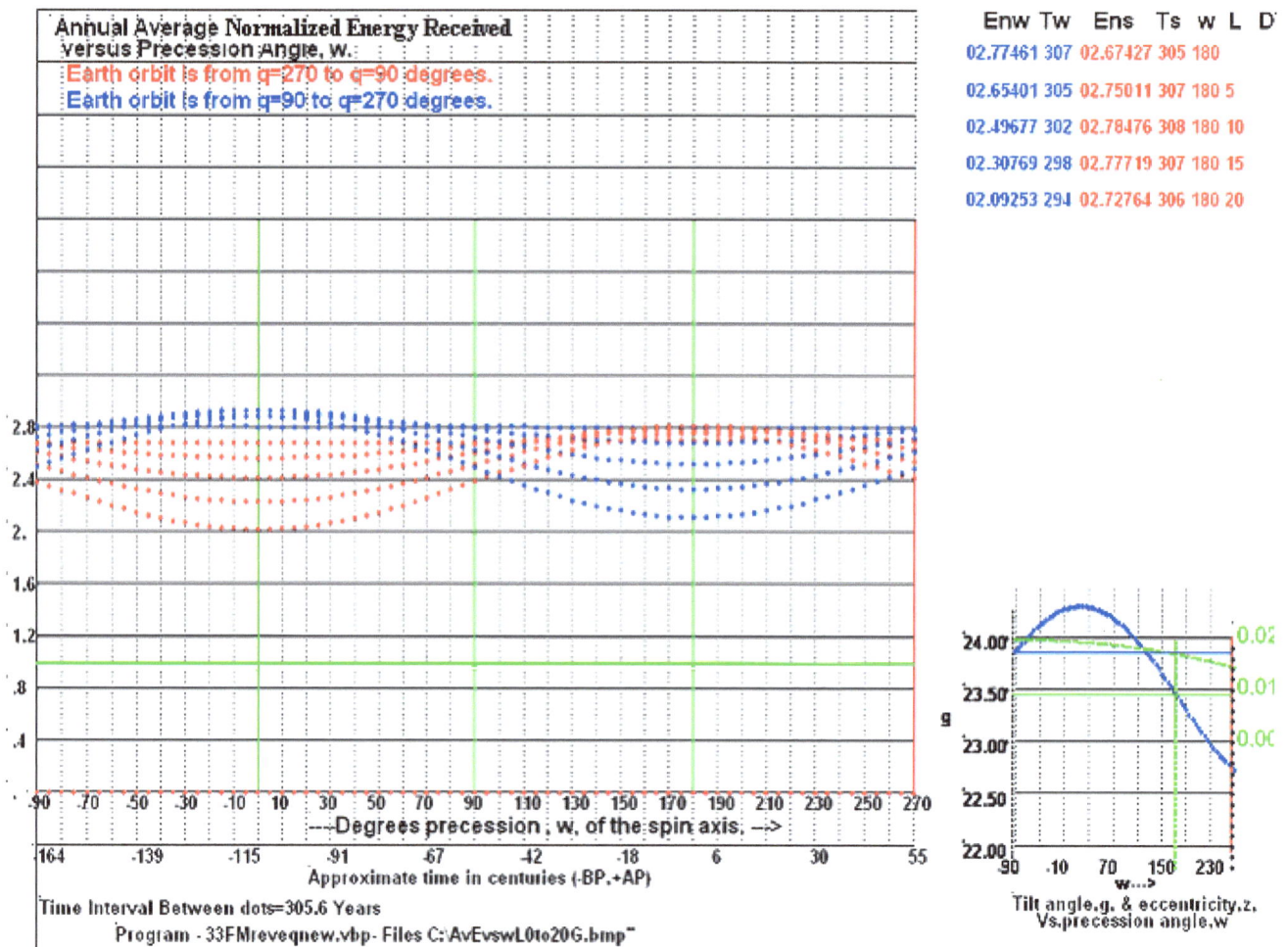

Figure(32)

Shows the energy received from the Sun (normalized to the energy received at w=180° at L=65° and for ω equal to 180°) at latitudes from 0 to 20 degrees.

In 1938 Milankovitch published calculations of what he called 'Summer Radiation', which is terminology for energy received during a small segment of Earth's orbit in the summer only. That portion of Earth's orbit used to calculate energy received in summer is unclear. It may have been for the month of July only. His calculations are shown in figure (33). Remember he was interested in predicting the onset of ice ages from 600,000 years before present and forward. Thus that portion in the first 22,000 years is the only part compared to the calculations in this monograph.

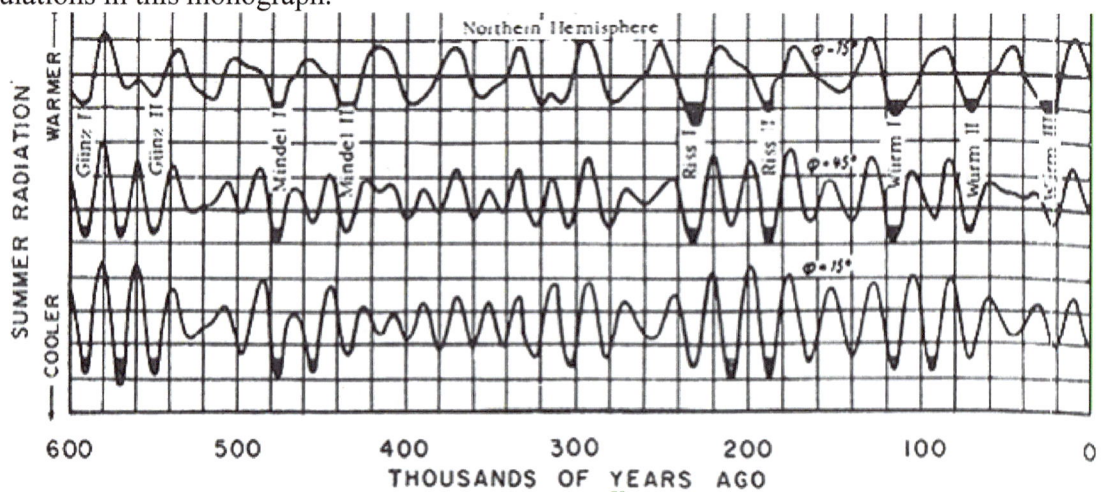

Figure (33)[58]

---

[58] From J.Imbrie and K.P. Imbrie , pg 108

The Milankovitch curves appear to differ in the range from 0 to 22,000 years past from those shown in figure (34). The Milankovitch curves predict the onset of glacier formation 22,000 years ago at L=75° and a cooling period starting about 10,000 years ago, which has not been observed in the Northern Hemisphere in America. The Great Lakes and Long Island Sound (latitude ~41°N) are the result of a warming period beginning about 9,500 years BP when the ice began to recede. Massive rocks were left on the Connecticut shore of Long Island as the ice receded. Long Island Sound and Long Island were created as the glacier pushed land in front of it. Figures (29) and (30) predict a warming period beginning about 9,500 years ago reaching a maximum warming currently or depending on the exact placement of the green line representing today, predicts the onset of a _cooling period_ in the near future or currently. Further, the Milankovitch curves predict glacier formation at 75° latitude between 20,000 to 30,000 years past and that melting should have begun 20,000 years ago. The current belief is that melting started 9, 500 years ago.

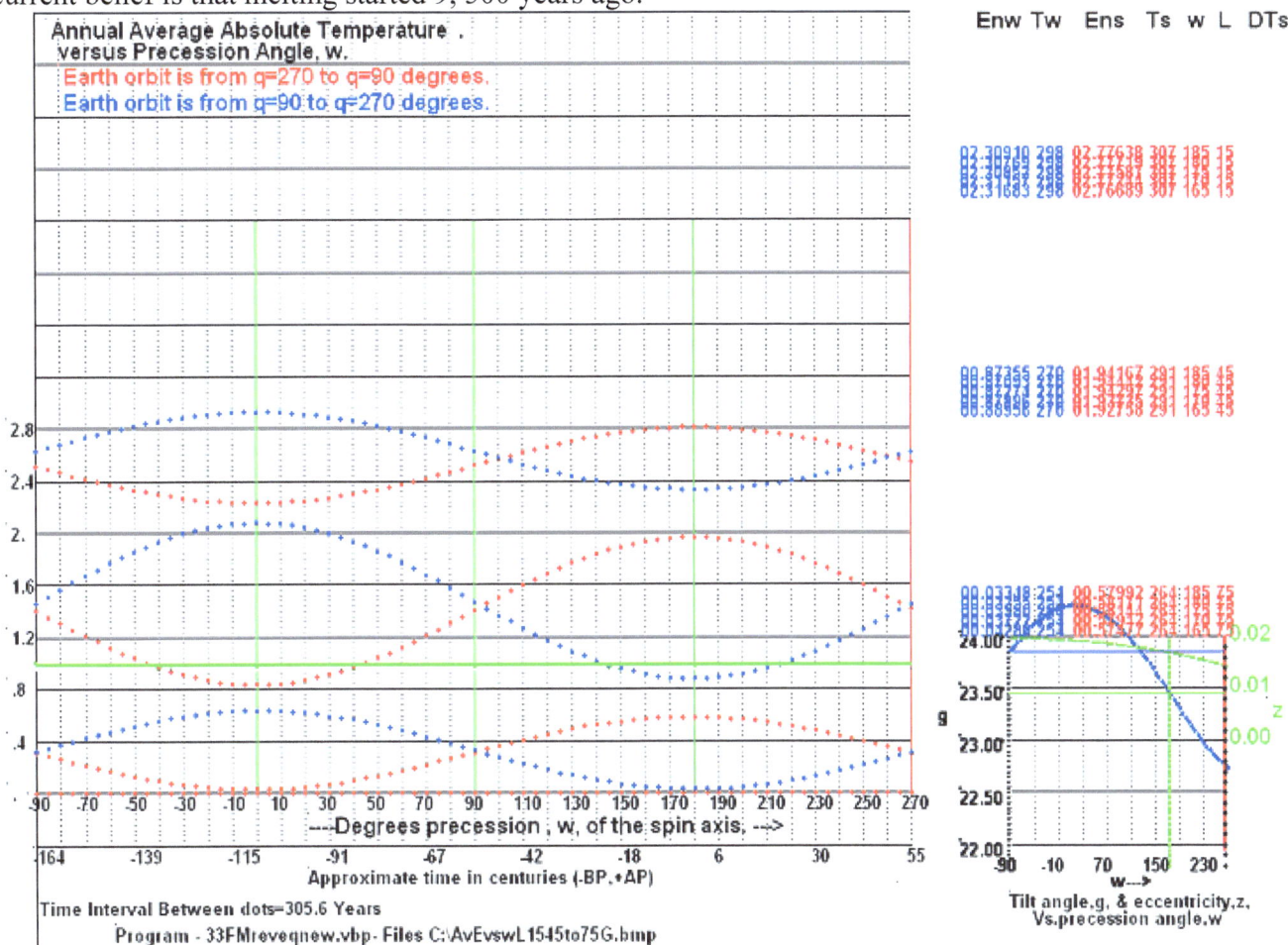

Figure (34)

Figure (34) shows energy received versus precession angle versus the same latitudes used by Milankovitch. The energy received is summed between Earth at θ=90° to 270° (blue curve) and 270° to 90° (red curve). The green vertical line at ω=180° is today and the ordinate, Summer Radiation, is similar to energy received, as in equation (17).

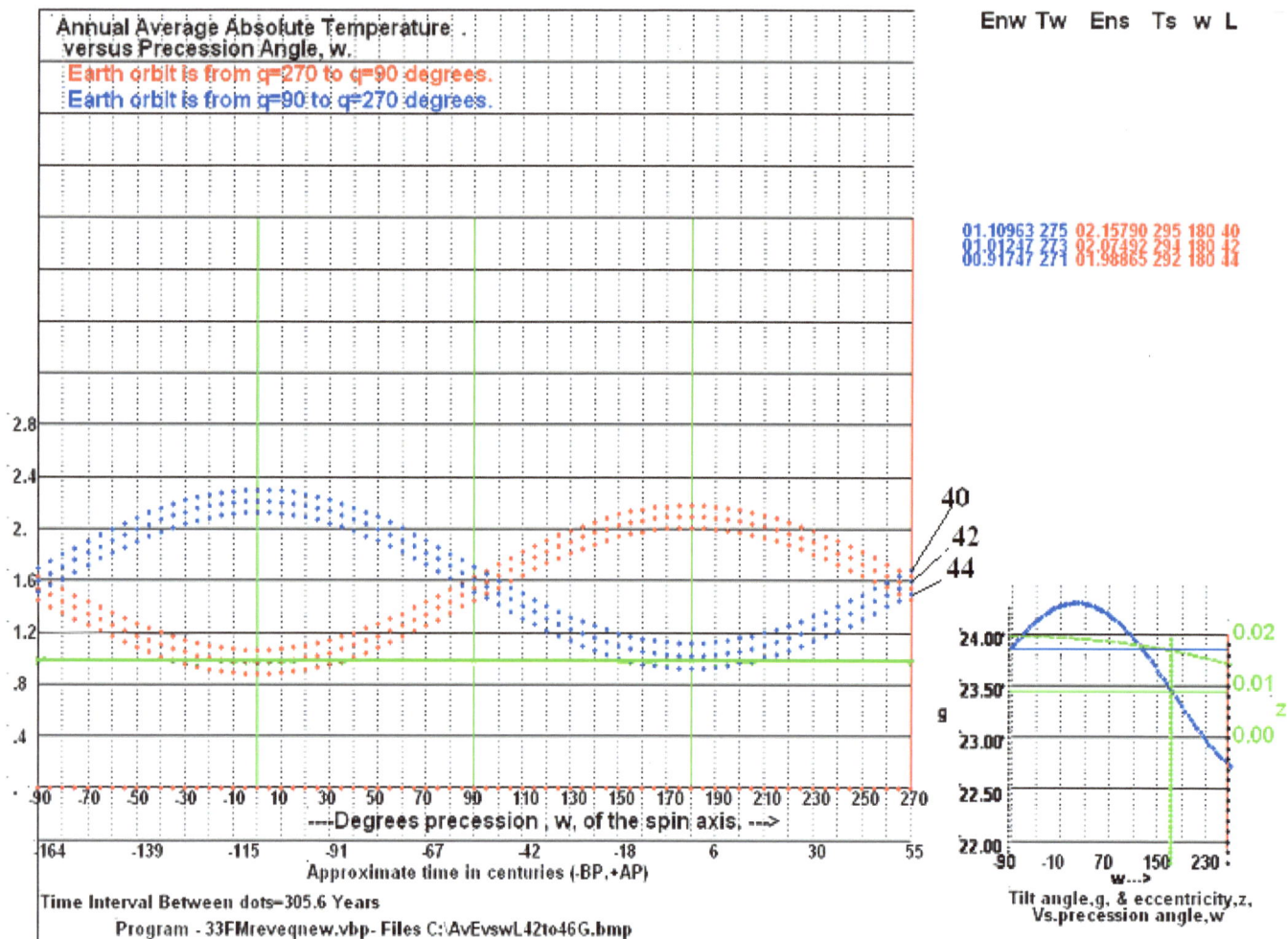

Annual Average Absolute Temperature versus Precession Angle, w.
Earth orbit is from q=270 to q=90 degrees.
Earth orbit is from q=90 to q=270 degrees.

Enw Tw Ens Ts w L

01.10963 275 02.15790 295 180 40
01.01217 273 02.07492 293 180 42
00.91747 271 01.98865 292 180 44

40
42
44

---Degrees precession , w, of the spin axis, --->
Approximate time in centuries (-BP,+AP)
Time Interval Between dots=305.6 Years
Program - 33FMreveqnew.vbp- Files C:\AvEvswL42to46G.bmp

Tilt angle,g, & eccentricity,z, Vs.precession angle,w

Figure (35)

Shows energy received vs. time for latitudes between 40 and 44 degrees. The green horizontal line is the normalized energy received at 65° North latitude and the vertical green line at ω=180° is the current time.

In figure (35) we see that for latitudes between 42° and 44° and for a time between approximately 91 to 120 centuries BP, the energy received in the 'winter season' drops below that received currently at 65° North latitude. A nominal current average temperature at 65° latitude is proximately 0°C. Note that the energy received in the 'red season' declines going forward, year after year. Hence, one could reason that *if temperature is a function of energy received*, at about 100 centuries ago the temperature at 44° latitude was below freezing and because the 'red season' temperature was declining, snow and ice was accumulating. However between 40° and 42° both red and blue temperatures were above freezing. The curves show clearly that around 42° the 'winter season' energy received by Earth started to increase at an accelerated rate. The logical consequence is melting of the ice sheets and the receding of the last glacier. In fact at about 95 centuries ago the ice sheets stopped advancing at approximately 41° and started to retreat. Evidently that process has been under way since 95 centuries ago. It is generally accepted that the glacier stopped advancing at Long Island Sound 95 centuries in the past. Further the curves predict that the receding of the ice sheets continues even today, but that the rate of melting is approaching zero and in the future will reverse leading to a return of the ice age.

To understand the consequences of increasing and decreasing energy received from the Sun we need an explicit equation that shows the functional relationship between incoming energy and temperature[60]. The

---

[60] For example the necessary condition for formation of snow or ice is that the temperature must be below 0° C

relationship between temperature and energy received could be derived using laws of physics or be an empirical relationship based on data obtained by direct measurement.

One of the most powerful concepts of physics is that of internal energy of a system. The internal energy is a function of measurable thermodynamic coordinates, such as absolute temperature, pressure, volume, entropy, and other intrinsic or extrinsic coordinates. Internal energy is described by an *equation of state* written in terms of measurable thermodynamic coordinates. Using the concept of internal energy and experimental data a useful relationship between energy received and temperature is derived below.

The Relationship between Energy Received and Absolute Surface Temperature.

The energy from the Sun that reaches Earth's surface is less than that at the beginning of the atmosphere. Some energy from the Sun is reflected back to space by particles in the air, clouds, and, some of the original radiation is reflected from Earth back to space.

The remaining energy is absorbed in the surface of Earth causing the surface temperature to rise. Some of the energy in the surface flows into Earth by thermal conduction and some flows into the air by thermal conduction and by convection(a mass transport process) and some of the surface energy is re-radiated at longer wave lengths (black body radiation). The later is absorbed by so called greenhouse molecules, for example CO_2 and CH_4, that are in low concentrations in the air boundary layer near Earth's Surface. Thus a layer of air in contact with the surface is heated by three processes; thermal conduction, convection, and long wave length radiation absorbed by green house molecules in the boundary layer. Hence, it is reasonable to expect that the resulting temperature of the air is a function of the energy received from the Sun.

The *internal energy* in the air layer near the surface is determined by or is a function of the energy that flows into the layer minus the energy returned to space.

Figure (36) shows measurements of the average surface temperature over sea, (A), and land, (B), of Earth as a function of latitude and time of the year[61].

Absolute temperature is an intrinsic thermodynamic variable that in the case of a gas is directly related to the energy of the molecules that make up the gas. The energy of air near the surface changes during one year because of pressure and temperature gradients, and, absorption of radiation. However, the annual average temperature is generally accepted as a measurement or indicator of energy in the air. Thus the temperature of the air near the surface of Earth averaged over a year is treated in this analysis as an equilibrium temperature.

The rate of change of temperature over water with latitude is less than that over land. Heat from the Sun absorbed by the ocean is dispersed from a given point or area more rapidly than by land because of convection in the water generated by wave action. In addition, if the water is in equilibrium with ice, the temperature is essentially constant and approximately 273°Kelvin[62]. Thus the rate of change of temperature as a function of latitude over water, particularly at high latitudes, is less rapid than on the surface of land[63].

The situation over land is very different (see figure 36B) where we see very large changes in temperature as latitude goes from 0° to 90°. According to figure (36B) the energy received in winter, in a band centered about the Equator, is greater than the energy received in the summer. This result is counter intuitive, but this result is predicted by equation (12) and the geometry underlying this result was discussed in detail in Chapter I, page 23. (Also see figure (31)). In this small detail, the theoretical energy received is consistent with experimental data.

At equilibrium the *internal energy* of the surface air layer, U, is a function of the absolute temperature of the air layer. Since climate, is basically determined by the temperature and motion of the air, we concentrate on the temperature distribution at or near the surface. Also note that energy captured by air is transported by

---

[61] Data in curve B is from Smithsonian Physical Tables 1964 and copied from the Encyclopedia Britannica.
[62] Salt in water reduces the freezing temperature. For simplicity freezing temperature of 273° is used throughout this analysis.
[63] Once ice is formed it is considered as land in this discussion.

mass transport, thermal conduction, and, radiation processes. Of the three processes, the later transports heat faster and further than the other processes. As discussed above, we know by direct observation that the local temperature can change rapidly over short periods of time (hours). To smooth out rapidly changing hourly temperatures, the average temperature versus latitude over a year is used to represent the steady state temperature.

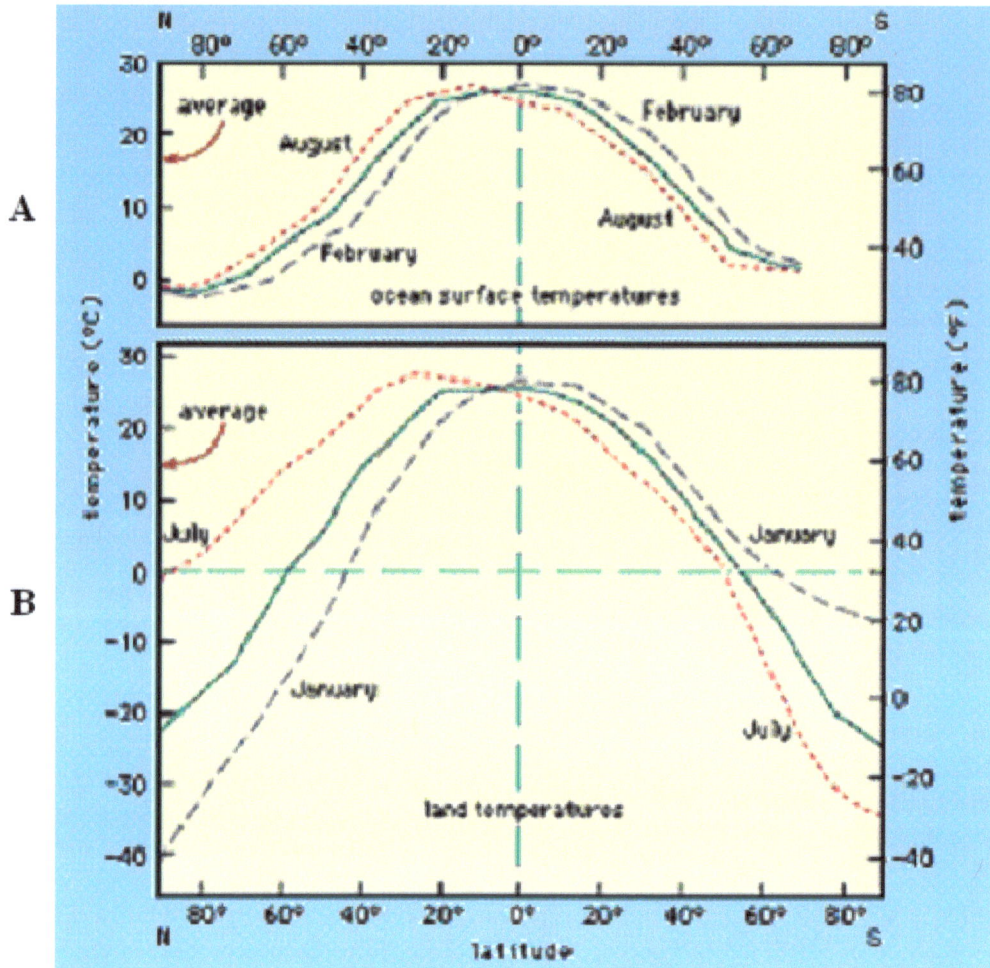

Figure (36)

Figure (36) shows average land and sea surface zonal temperatures for a year as a function of latitude and season in the Northern Hemisphere[64]. The red curve (*see figure (36B)*) shows the average temperature in July, the warmest month, and the black curve shows the average temperature in February, the coldest month. The green curve shows the average annual temperature versus latitude. Notice that in figures (36A) and (36B) the temperature at the Equator is higher in winter than in summer. According to the theory, (*see Chapter I, equation (12)*, and, *page 30)* the energy received in winter at the Equator is greater than the energy received in summer, thus the theory is in agreement with experiment. The average annual temperatures at airports in the US[65] and remote places on Earth, agree with average annual temperature data shown in figure (36)[66].

Average Temperature as a Function of Energy Received.

The annual average temperature and annual average energy received are both functions of latitude, hence, the temperature as a function of energy received is readily obtained. The average zonal temperature and zonal

---

[64] Ocean: Average Temperature, Encyclopedia Britannica 2006. The primary source of this data is from G.Wust, W.Brogmus, and, E Noodt, Kieier Meeresforschungen,vol.10 (1954) ,(top), (bottom), data from Smithsonian Physical Tables (1964).
[65] San Francisco, CA, Chicago, Il, New York, NY, Beaufort SC, North West territories, Canada, 70.5°North latitude, Arctic Ocean, 80.6°N
[66] See The Cause of Global Warming, Vincent Gray, Lecture to the Wellington Branch of the Royal Society of New Zeeland, 22nd November, 2000, Fig.(15) and (16). Also available at http://www.john-daly.com/cause/cause.htm

energy received versus latitude are obtained by determining the temperature and energy received at variously spaced longitudes around the Earth. The program used to calculate energy versus latitude sums energy received over a small area for each, say, 1° increment in longitude versus latitude. Thus the program calculates zonal energy received versus latitude. Using the energy vs. latitude and temperature vs. latitude to replace latitude, the average temperature versus average normalized theoretical energy is obtained. The average temperature versus average normalized theoretical energy is shown in figure (37).

Clearly, ***the annual average zonal surface temperature over land is a linear function of the theoretical average energy received. In addition, the directly measured energy received at the surface is also a linear function of theoretical energy received[67].***

This useful relationship between zonal temperature and energy received provides a way to estimate the change in surface temperature per century, a quantity frequently reported in the literature.

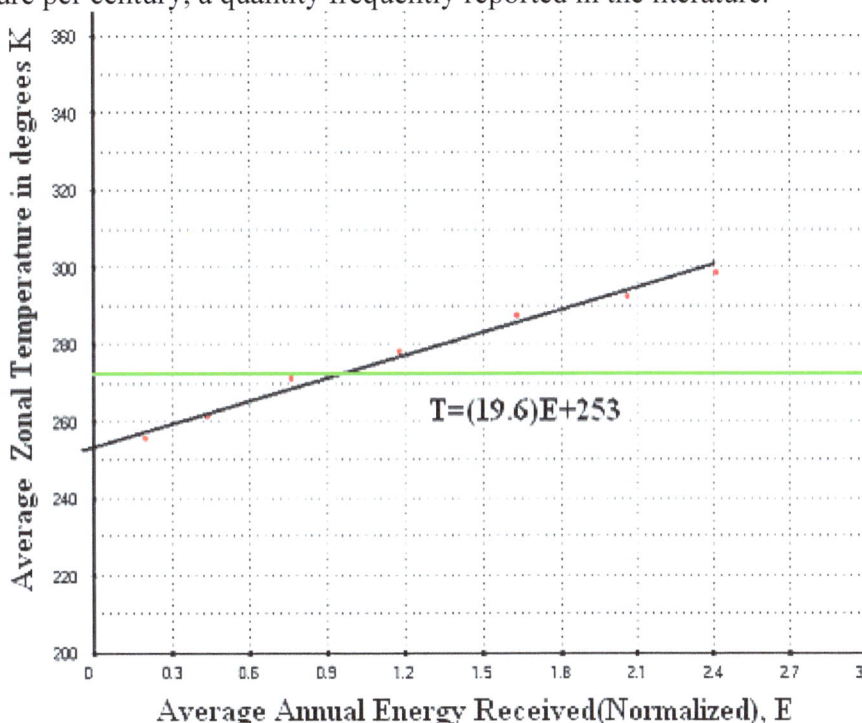

The figure shows a plot with y-axis "Average Zonal Temperature in degrees K" ranging from 200 to 360, and x-axis "Average Annual Energy Received(Normalized), E" ranging from 0 to 3. The equation shown is:

$$T = (19.6)E + 253$$

Figure (37)

Figure (37) shows annual *average* surface temperature, in degrees Kelvin, over land in the Northern Hemisphere as function of the annual average energy received obtained using the program to calculate E_{band} ($\varepsilon, \gamma, \theta, \omega, L$). The average surface temperature over land is obtained from figure (36B) and the energy data is obtained from figure (29). *The relationship between zonal surface temperature obtained by direct measurement and the zonal theoretical energy received* is linear. At temperatures below the green line, 273^oK, water becomes ice, a solid. Above the line water is a liquid at normal atmosphere pressure.

It is important to emphasize that the graph in figure (37) is an *empirical* surface air temperature to energy received at the limit of Earth's atmosphere relationship. As expected the annual average zonal temperature is a function of the energy received at that zone. The statement is a universal statement. It can be viewed as an ***equation of state*** written in terms of two thermodynamic coordinates, namely temperature, T, and energy, E. It is a steady state equation based on data obtained from experiment and applicable only to a steady state condition. The temperature as shown in figure (37) has been obtained by measuring the temperature many times during the day and night to obtain an average temperature. Thus the averaging process yields a

---

[67] Verified using data from The Hand Book of Chemistry and Physics, 57th Edition page F201.
[70] Ibid, F. Singer, Figure (21) page 21

temperature that approximates the steady state value of temperature. A similar procedure is used by researchers reporting global temperature trends. Usually, the temperature is the temperature averaged over the preceding five years. Since the calculated energy received is an average of daily energy received measured over (364)/2 days, the calculated energy average over 182 days corresponds to or is equivalent to a steady state value of zonal energy received. In addition, data shown in figure (7) Chapter I, further supports the result that temperature is a function of latitude.

One would expect that if no energy is received, the temperature should be zero. In fact, it is not zero but equal 253° Kelvin. What then is the source or cause of this residual temperature? We know that Earth retains some of the incoming radiation that reaches the surface and that it returns some of that energy as black body radiation. Molecules in the atmosphere such as H_2O, CO_2, and, CH_4 can absorb some of the black body radiation thereby increasing the internal energy of the atmosphere in which they are dissolved. Since temperature is directly related to internal energy, the temperature of the air, of which H_2O, CO_2, and, CH_4 are a part, increases.

Ultimately, a steady state temperature distribution in a layer near Earth's surface is reached because the air not only can receive heat from Earth by conduction and convection it also gains heat from black body radiation from the surface. If the GHG's absorb black radiation, they also emit that radiation back to space in all directions. In addition, a given point on Earth sends black body radiation in all directions relative to the surface as do all neighboring points. Because of the higher rate of energy transfer by radiation most of the heat transfer is by radiation. In the steady state, one should then expect a gradient in temperature normal to Earth and a lesser gradient parallel to Earth's surface. Thus at steady state a 'blanket' of heated air is created near the surface of Earth. Apparently, the residual temperature (the intercept in figure (37)), is a manifestation of this additional energy. Further the intercept appears to be **_independent_** of latitude. Why is it constant and independent of latitude? What then is the explanation for this curious result? The energy absorbed at the surface drives the generation of infrared radiation. The energy received decreases as does the IR emission from Earth as latitude increases thus one should expect the temperature of the 'blanket' to decrease at high latitudes. Since measurements of the concentration of CO_2 show only about a 6% increase between latitudes 0° and 90°,some thing else is the cause[70]. Could it be that the uniform source of energy at the surface is from the molten core of Earth?[71] We know from mines and drill holes that, near the surface of the Earth, the temperature increases by about 1 degree Fahrenheit for every 60 feet in depth.
Nevertheless equation (20) appears to predict correctly observed current and BP temperatures (See figures (39) and (40)). Further discussion of the 'temperature blanket' including the independence of temperature on *latitude* and GHG *concentration* is developed in Chapter III.

Estimating the Temperature and Rate of Change of Temperature.
Referring to figure (37), we see that a linear relationship can be used to relate surface temperature and the theoretical average energy received, E. Since the internal energy, U, of the air is a linear function of temperature, we can write[73],

$$U=CT,$$

where, C is a constant and T is the absolute temperature. A change in internal energy is,

$$\Delta U=C(T+\Delta T)- C(T)=C\,\Delta T.$$

The fractional change in internal energy for an input of, ΔU, is,

$$\Delta U/\, U = \Delta T/T.$$

Since a change in internal energy of the air layer near the surface is proportional to the incoming energy[74], E, from the Sun we can write,

---

[71] We know from mines and drill holes that, near the surface of the Earth, the temperature increases by about 1 degree Fahrenheit for every 60 feet in depth.
[73] See H.B. Callen, Thermodynamics, John Wiley&Sons,1962, Appendix D, page 335
[74] See footnote 67

$$\Delta E/E = \Delta U/U.$$

Hence, we can obtain temperature changes because,

$$\Delta E/E = \Delta U/U = \Delta T/T, \text{ or,}$$
$$\Delta T = T(\Delta E/E)_{L,\omega}.$$

.............eq.(19)

The absolute temperature, T, and, $(\Delta E/E)_{L\omega}$, are evaluated at a given latitude and precession angle. Referring to figure (37) the absolute temperature is given by,

$$T = (19.6)E + 253. \qquad .... \text{ eq. (20)}$$

The data in Tables (1) and (2) is from tables shown in figures (29) and (30) in columns En, ω, and, L, the normalized energy received, the precession angle, and, the latitude, respectively.

Using the quantities, $\Delta E/E$ from the tables and the precession angle, the rate of change of temperature per century as a function of ω is obtained and tabulated in Tables (1) and (2))[75]. Tables (1) and (2) list the temperature and the average increase or decrease in temperature per century as a function of latitude and the precession angle as Earth moves in orbit from θ equals $-90°$ to $90°$. Since the precession angles used are between $\omega = 170°$ to $185°$ the rates of temperature change range between 611 years BP to 306 years AP and show recent rates of temperature change in the summer when snow pact and glacier melting is observed.

A Sample Calculation of Rate of Change of Temperature per Century.

Using the data in figure (29) and Table (1), the average temperature in the summer ($\omega=180°$) at $30°$ North latitude is $302°$K. The fractional change in energy received in the last 305.6 years is $\Delta E/E =$ (2.50989-2.50832)/ 2.50832 = (0.00157)2.50832=0.000626 units. Using Δt=305.6years, and equation (19) the rate of change of temperature at 30 degrees latitude, in degrees Kelvin per century in the summer is, =0.000626 *100/306.5*302 or, $0.062°$ C/Century at $30°$ latitude or $0.06°$C/century after rounding.

Past and Future Temperature Changes.

The temperature change per century for L=$15°$ to $85°$ for the past and future are shown in Tables (1) and (2). At the current precession angle, $\omega=180°$, the average rate of temperature change is 0.06 C$°$/century. For latitudes between $15°$ and $80°$ and current time ($\omega=180°$) the rate of temperature ranges from 0.052 to 0.035 C$°$/century. The current maximum temperature change is 0.14C$°$/century at $75°$ latitude. Nine hundred years earlier the rate of temperate rise at $75°$ latitude was 0.65C$°$/century. Thus less rapid melting of arctic ice is predicted in the current summers than in prior years. If not now or in the immediate future years, the rate of temperature change goes negative leading to growth and subsequent advancement toward the south of a snow-ice sheet. Further, note that for times 912 years BP and latitudes between $60°$ to $80°$ latitude the predicted average rate of increase in temperature was 0.38 to 0.66 C$°$/Century, respectively, (See figure (38)).

Table (1)
Past, Present, and, Future Rates of Warming or Cooling.

latitude in degrees	$\Delta\omega$	Temperature degrees K	Rate in C$°$/century BP	Rate in C$°$/century AP
15	185-180	307		-0.03
15	180-175	307	+0.05	
15	175-170	307	+0.12	
15	170-165	307	+0.20	
25	185-180			-0.06
25	180-175	305	+0.06	
25	175-170	305	+0.18	
25	170-165	305	+0.30	
35	185-180	299		-0.10
35	180-175	299	+0.07	
35	175-170	299	+0.33	

---

[75] An increment of $5°$ in the precession angle corresponds to a time interval of 305.6 years.

latitude in degrees	$\Delta\omega$	Temperature degrees K	Rate in C°/century BP	Rate in C°/century AP
35	170-165	299	+0.39	
45	185-180	291		-0.13
45	180-175	291	+0.073	
45	175-170	291	+0.28	
45	170-165	291	+0.48	
55	185-180	282		-0.17
55	180-175	282	+0.073	
55	175-170	282	+0.32	
55	170-165	282	+0.56	
65	185-180	273		-0.22
65	180-175	273	+0.064	
65	175-170	273	+0.35	
65	170-165	273	+0.635	
75	185-180	264		-0.26
75	180-175	264	+0.14	
75	175-170	264	+0.35	
75	170-165	264	+0.65	

Table (2)

latitude in degrees	$\Delta\omega$	Temperature degrees K	Rate in C°/century BP	Rate in C°/century AP
20	185-180	306		-0.047?
20	180-175	306	+0.052	
20	175-170	306	+0.153	
20	170-165	306	+0.253	
30	185-180	302		-0.085
30	180-175	302	+0.062	
30	175-170	302	+0.205	
30	170-165	302	+0.349	
40	185-180	295		-0.116
40	180-175	295	+0.068	
40	175-170	295	+0.254	
40	170-165	295	+0.439	
50	185-180	287		-0.153
50	180-175	287	+0.073	
50	175-170	287	+0.297	
50	170-165	287	+0.524	
60	185-180	277		-0.195
60	180-175	277	+0.078	
60	175-170	277	+0.380	
60	170-165	277	+0.604	
70	185-180	268		-0.251
70	180-175	268	+0.048	
70	175-170	268	+0.345	
70	170-165	268	+0.646	
80	185-180	261		-0.275
80	180-175	261	+0.036	
80	175-170	261	+0.34	
80	170-165	261	+0.66	

Tables (1) and (2) list the theoretical temperature and the rate of temperature change per century for latitudes between 15 and 80 degrees and changes in ω from 165° to 185° or from 9 BP to 3 centuries AP.

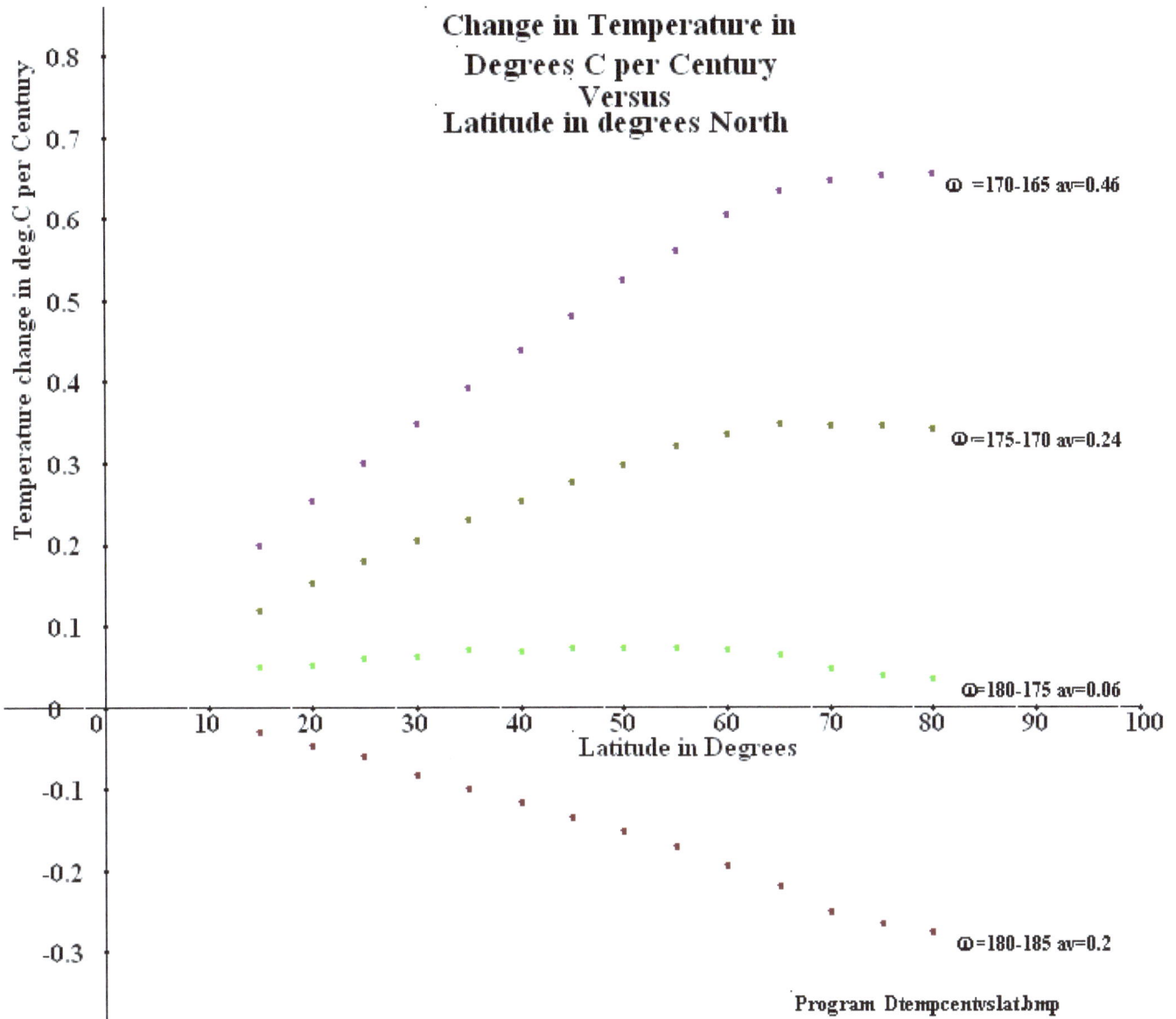

Figure (38)

Figure (38) shows the rate of change of temperature as Earth moves in its orbit from $\theta = -90^\circ$ to 90° as a function of latitude for recent values of the precession angle, ω. Note that the maximum rate of increase in temperature is above 50 degrees latitude. In the future, the rate of change is negative. Hence in the immediate future one should expect onset of snow and ice accumulation at the higher latitudes because the summer temperatures will be declining and the winter temperatures will be below freezing. (See figure (39) below.)

From figure (39) we see that conditions for ice formation begin at latitudes greater than 65° N. latitude. Currently, the rate of summer temperature change is negative and trending more negative going forward. Prior to current times the rate of temperature change at latitudes between 50° and 70° was between 0.48°/century and 0.65°/century and positive. Thus we may have reached a peak in summer temperature and should expect a progressive decline in temperature going forward.

Chapter III Comparison of Theoretical Results with Measurements.

Measured Versus Theoretical Rates of Change.

From Table (2) we see that the predicted rate of change of energy received per century is a function of the latitude. Thus when reporting measurements of current temperature change per century one must specify the latitude. Some direct measurements of surface temperature in the literature are a compilation of average temperatures changes over a range of latitudes[76]. In addition, the reliability of surface temperature measurements has been challenged by recent investigators[77,78]. The theoretical results in this monograph are calculated at specific latitudes. In addition, the predicted rates of change increase at the higher latitudes, especially, above 65° North latitude. Consequently, a precise quantitative comparison of measured temperature change per century to the theory with some data is compromised. However, in some literature the temperature versus time is reported for specific latitudes (See footnotes 61, and, 63, and, figure (36).

Using the theoretical rate of change of temperature for the last 306 years averaged over latitudes from 15 to 85 degrees yields 0.06°C per century. Between 306 and 611 years BP the theoretical rate of change of temperature is 0.26°C per century and between 611 and 917 years BP, 0.46° per century.

The theoretical rate of change of temperature averaged over latitudes from 15 to 85 degrees for the last 100 years is 0.06°C per century. This value appears to be in the same range as global lower troposphere temperature obtained from satellite data.[79]

Average temperatures measured in remote parts of the world by instrumentation carefully constructed according specification and sited to avoid nearby heat sources, have been constant for the last 30 to 40 years[80].

Recent temperature measurements obtained by different methods in the Northern Hemisphere indicate a near zero rate of temperature change per century. The same temperatures and rates of change of temperature are predicted by the theory developed in this monograph.

In some time periods the theory does not predict the temperature and trends because of **unpredictable events** such as volcanic eruptions, meteorite and comet, collisions with Earth that place particles in the atmosphere that diminish the energy incident on Earth's surface.

Unpredictable Events

Unpredictable vents, such as meteorite and comet collisions and volcanic eruptions also initiate temperature change because they cause dust in the atmosphere that reduces the energy received from the Sun. These unpredictable events appear to always cause the temperature to decrease and last for times that depend on the magnitude of the energy associated with the event.

Such events often extend the formation of ice sheets to lower altitudes and usually persist for as long as 10 centuries. *Unpredictable events* in the prior 200 centuries have caused Earth's temperature to drop rapidly. One such event, the Younger-Dryas, occurred 130 to 116 centuries BP and lowered the theoretical temperature distribution pattern prior to the beginning of the retreat of the last ice age.

In addition, changes in the radiant output of the Sun, for example those caused by the periodic sun spot cycle, can superimpose significant changes in the temperature[81]. These unpredictable events are superimposed on the predictable quasi-periodic events caused by the predictable motion of Earth and the direction and possibly the magnitude of the spin axis vector.

---

[76] See footnote 79

[77] Anthony Watts, Is the U.S. Surface Temperature Record Reliable? , Chicago, Il: Heartland Institute, 2009

[78] Vincent Gray, Lecture to the Wellington Branch of the Royal Society of New Zealand, , 2000, also see htpp://www.john-daly.com/cause/cause.htm,2009

[79] S. Fred Singer, ed, *Nature, Not Human Activity, Rule the Climate Report of the NIPCC*, Heartland Institute, pg.,10, Figure(13)

[80] Ibid, Vincent Gray, Figures, 11, 12, 13, 14,,15, 16,17,and, 20,2000.

[81] The energy radiated from the Sun is modulated by the appearance of a 11 year sun spot cycle and associated cosmic rays that cause an increase in cloud cover, hence reducing energy received .

The combination of predictable and unpredictable events determines the energy received and therefore the history of the temperature distribution on Earth throughout time.

The Little Ice Age that appeared between 15 to 19 centuries BP was probably caused by a sequence of unpredictable events superimposed on a predictable trend.

Theoretical Absolute Temperature versus Precession Angle.

Using equation (20) the coordinate, labeled, 'Energy Received', in figures (29) and (30) is replaced by the corresponding temperature in figures (39) and (40). If the temperature is below 273° K the points are plotted in green.

The temperatures are obtained assuming that only the empirical equation, equation (20), the direction of the precession angle, the tilt angle, and, the orbital position of Earth, absent unpredictable events, determine the distribution of Earth's surface air temperature.

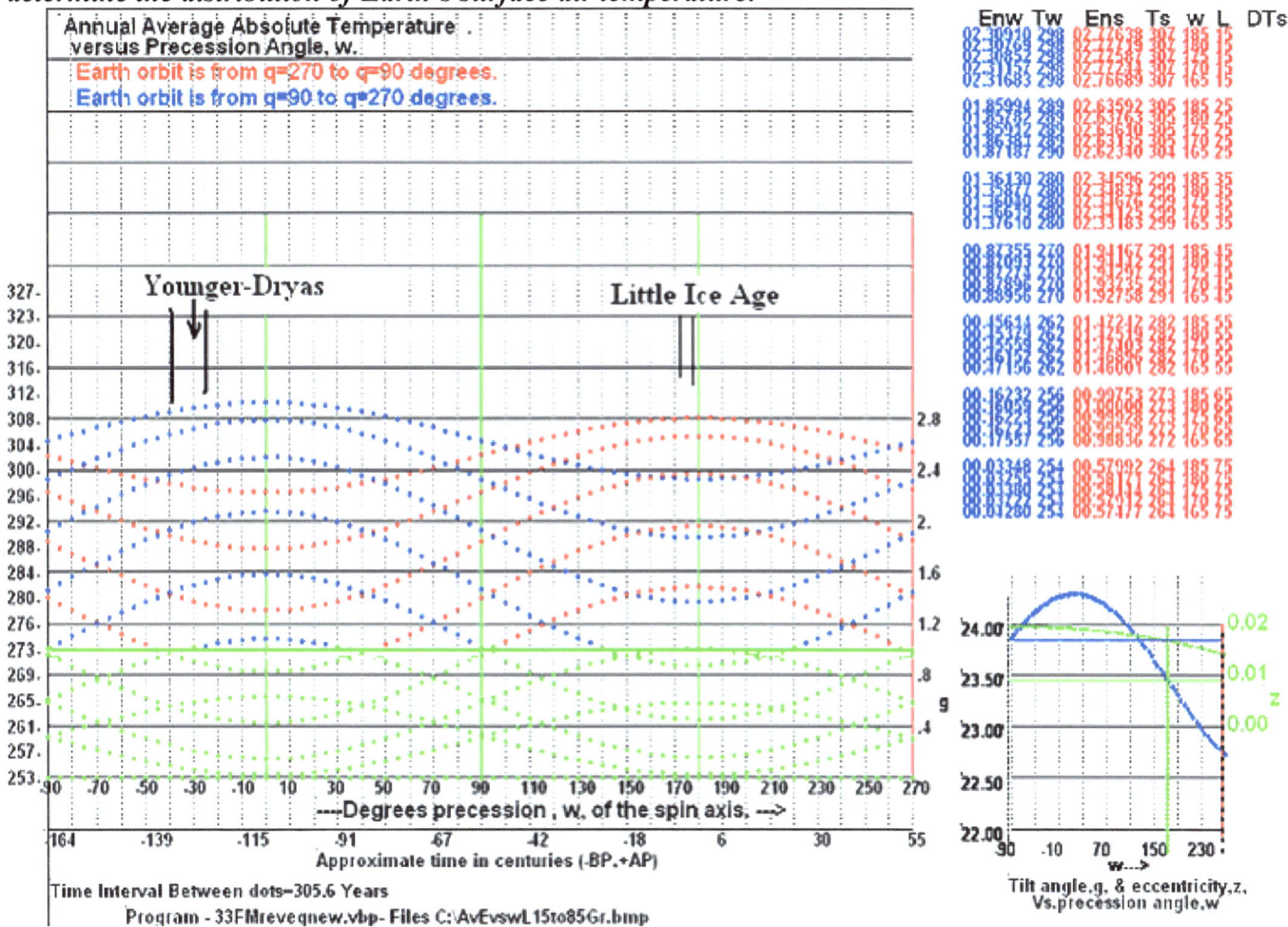

Figure (39)

Shows the average temperature in degrees Kelvin as Earth travels from $\theta=270^\circ$ to 90° (red) and from $\theta=90^\circ$ to 270° (blue) for latitudes from 15 to 75 degrees. If the color is green the average annual temperature is below 273°K. The normalized energy received is shown on the right. The time and duration of the unpredictable events, the Younger-Dryas event between 135 to 125 centuries the Little Ice Age between 1500 to 1850 years BP are indicated.

Interpreting the Temperature versus Precession Angle Curves.

From cursory examination of the temperature versus precession angle curves in figure (39) we see that at about 110 centuries BP or $\omega=0^\circ$, the summers were warmer and that the winters were colder than today. Further, going forward from $\omega=0^\circ$, the summer temperature *decreases* but as ω approaches 180° (today), the temperature is ***increasing***. Going forward from ω=0°, the winter temperature ***increases*** then as ω approaches 180° the winter temperature is ***decreasing***. Clearly, the curves show absolute temperature, trends in temperature, and, latitudes at which water freezes, all as a function of ω or time BP. Since the onset and rate of retreat of glaciers, absent unpredictable events, is clearly a function of temperature and temperature trends we can use the curves to predict the onset and rate of retreat of glaciers. But first let us study the conditions that lead to snow fall, glacier formation, and, glacier retreat.

Glacier Growth and Retreat

Note in figure (39) that at 89 centuries BP and L=45° latitude, the temperature was just transiting from freezing to melting in winter. Thus the snow and ice pack that had been accumulating in prior years because of falling summer temperatures, begins to melt marking the initial retreat of the ice pack (glacier) at about 45° latitude at about 89 centuries BP. The temperature in the summer and winter never drops below the freezing point until approximately 12 centuries BP (ω=150°), when winter snow appears. The snow never accumulates to form an ice/snow pact because the summer temperature increases each year going forward from ω=150° insuring melting before more snow is transported in winter, hence negating snow accumulation. Prior to the point of onset of melting (ω<40°) the snow and pack ice had been accumulating because the summer temperature was declining and winter temperature was always below freezing. Note that the theory predicts that glaciers should still exist at the surface of Earth (*sea level*) above 65° latitude today and they do because summer and winter temperatures are both below freezing. In addition, in the future the 'summer' temperature is declining and the winter temperature is below freezing, suggesting that if a mechanism exists for transport of equal or greater amounts of water to latitudes above 65° latitude, snow and ice should accumulate thus creating glaciers in the future.

To gain further understanding of how, why and when glaciers grow consider the curves shown in figure (40).

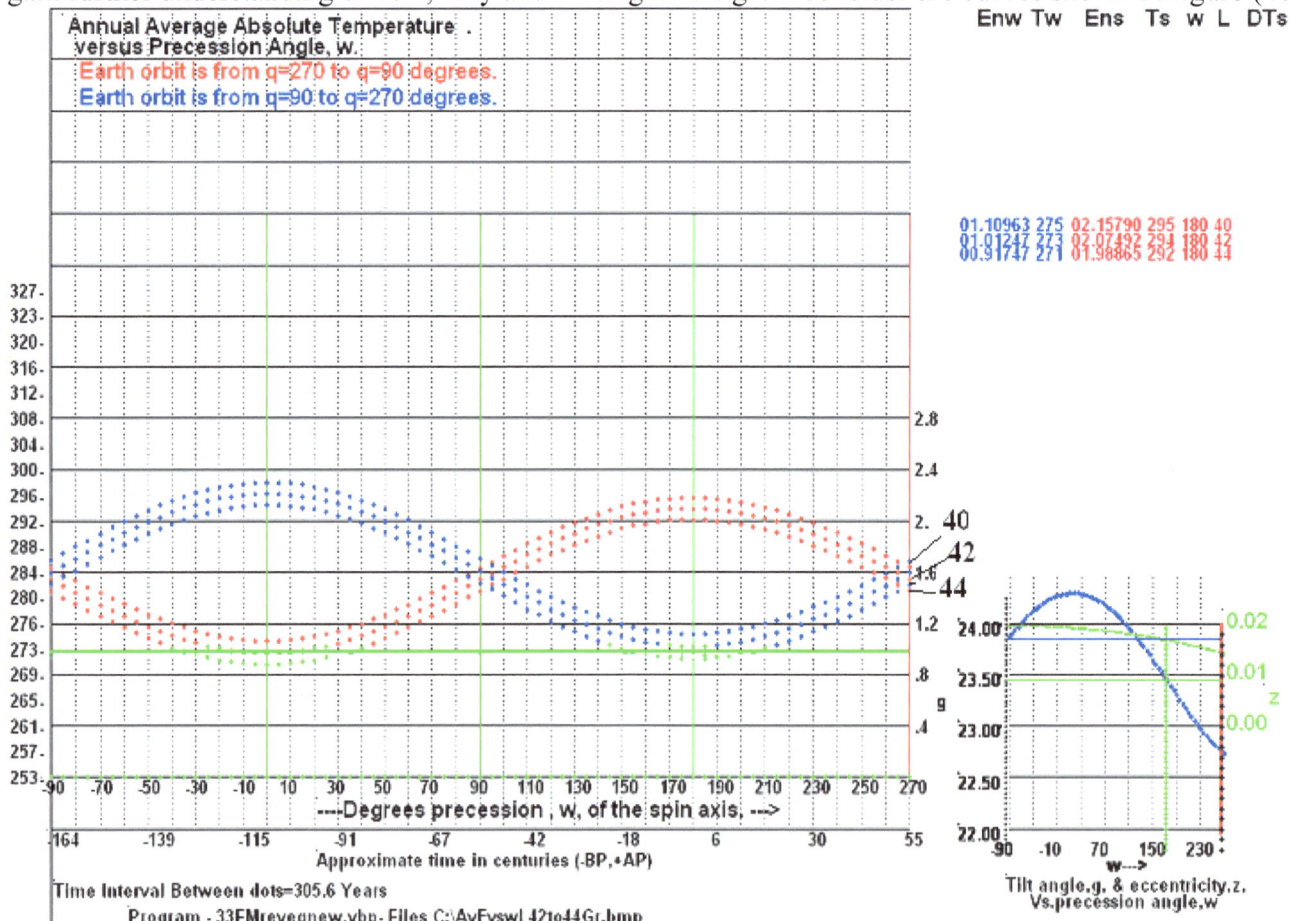

Figure (40)

50

Figure (40) shows average temperature versus precession angle for latitudes L=40°, 42°, and, 44° as Earth travels in orbit between θ equals 270° and 90° (red) and between 90° and 270° (blue). The numbers at upper right show the energy received and corresponding latitude and average winter and summer temperature at w=180°.

At L=40° and for ω between -10° and +10° the annual average temperature is never below 273°, hence, no snow can form and if snow exist, caused by an unpredictable event, it will ultimately melt. The temperature at all lesser latitudes never drops below 273° at ω=0°. Therefore snow never forms at latitudes below about 40°. However, between ω=170° and 180° latitudes between 42° and 44° snow falls in the winters but melts in the summer.

At latitude=44° and for ω between -10° and 0° the winter temperature is below 273° and during the summer part of the annual cycle above 273° and increasing year after year. Hence, snow forms, but the summer temperature is increasing each year causing the prior winter snow to melt. Between ω=0 and 10° the summer temperature is decreasing and winter temperature is below freezing. As a consequence winter snow accumulates year after year causing glacier growth. At 42° (the middle curve) and for ω between 0° and +10° the winter temperature is only slightly below 273°, hence, ice can form but since the temperature is above 273° going forward all of the ice formed will melt.

The later case represents the onset of glacier retreat and begins at 95 ± 3centuries BP. It is well documented that the North American continent began warming about 9500 years BP and the ice began to recede from about 41° North latitude. The last Ice Age glacier advanced to Long Island Sound (~ 41° latitude) stopped and began to retreat.

As seen in figure (39), at lower latitudes, ice could not form because the temperature never dropped below 273° K.

Declining Future Temperature.
In addition, the theory predicts that, now or beginning soon, a ***cooling period,*** lasting several thousand years, will begin. Tables (1) and (2) and figure (39) show that for all latitudes greater than about 15 degrees, the rate that energy received in the summer prior to today has been ***increasing***, but in the immediate future the energy received at latitudes greater than 15° will be ***declining each year.***

Figures (38), (39), and, (40) show that the ***current*** (see ω=180°, the green vertical line) summer temperature at all latitudes are at or are approaching a broad and nearly flat maximum and then decreasing going forward to the future. Therefore, one should expect the average summer temperature in the future throughout the Northern Hemisphere to be constant or decline. This conclusion appears to agree with recent satellite data[82] and high US temperatures recorded since 1880[83]. Winter temperatures in the future for latitudes greater than 40° latitude are all below freezing. Since summer temperatures are decreasing, one should expect in the future, snow to accumulate, starting at the higher latitudes, leading to onset of glacier growth.

The Rate of Warming and Cooling.
The predicted rate of warming per century is based solely on empirical equation (20), derived from Northern Hemisphere average annual temperatures recorded in the mid 1960's, changes in Earth's spin axis angle, the advance of the precession angle, and, changes in the orbit of Earth. The predicted rate of current warming is of the order of magnitude of +0.06 C°/century. Clearly this is less than that reported in *some* recent literature.[84]. However, the reliability of the data sets used by to afore mentioned researchers is doubtful.[86] In addition, other

[82] S. Fred Singer, ed, *Nature, Not Human Activity, Rule the Climate Report of the NIPCC*, Heartland Institute, pg.,10
[83] Ibid., Figure(13), pg. (25), Figure (25)
[84] J. Hanson, R. Ruedy, Miki, et al, NASA and Columbia Univ. Earth Institute, Global Temperature Trends:2005 Summation, 2005, Figure 1
[86] Anthony Watts, Is the U.S. Surface Temperature Record Reliable? Chicago IL: The Heartland Institute, 2009

research (See figures below) shows that in recent years the temperature in the USA has been constant or declining.[87]

Annual mean Surface temperature record in °C for Clyde, Northern Territories, Canada

70.5N 68.5W   1943-1999

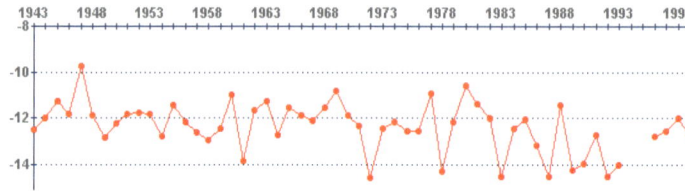

Surface record for Franz Josef Land, Arctic Ocean

80.6N 58.0E

Northern Siberia Stations
Annual Mean Temperature [°C]

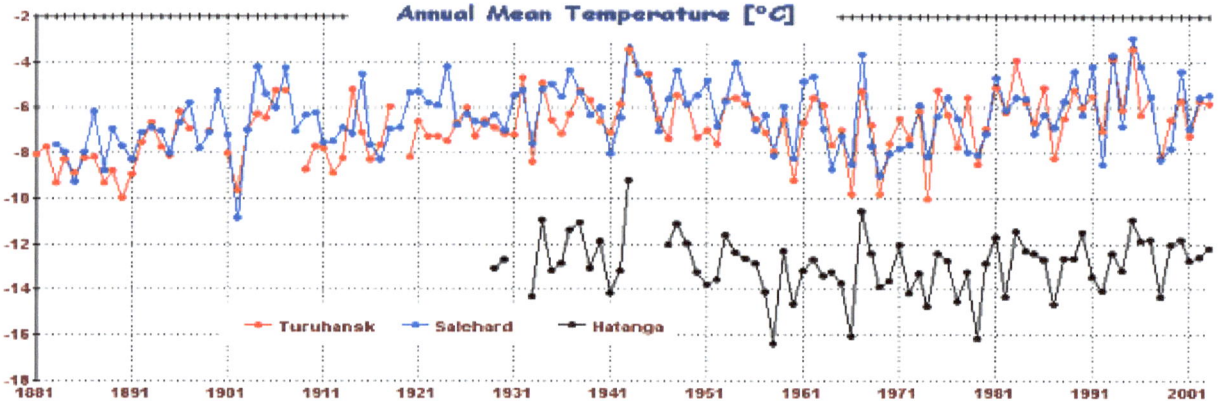

Godthaab & Angmagssalik, Greenland
Annual Mean Temperature [°C] 1866-2003

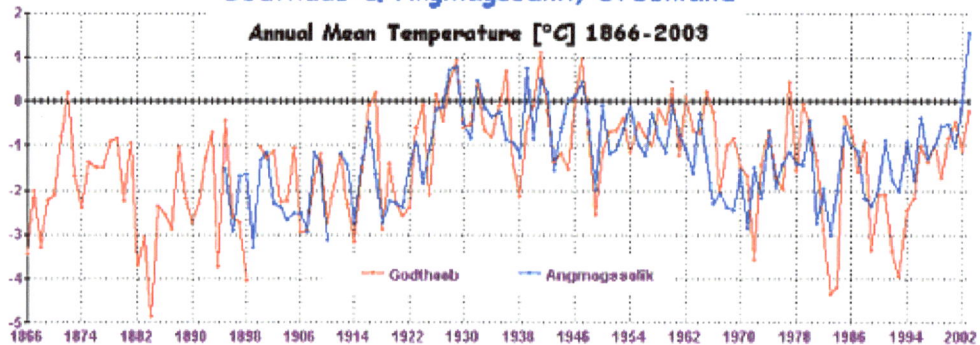

Figures (41)

Figures (41) are from a Lecture to the Wellington Branch of the Royal Society of New Zealand by Vincent Gray, 2000 and show temperature in degrees Centigrade vs, time in remote areas away from densely developed regions.

---

[87] Ibid.,Vincent Gray,   Figures,15, 16,17,and, 20,2000.

The predicted temperatures and rates of change of temperature are in good agreement with recorded temperatures for corresponding latitudes and time intervals.

In addition, the predicted rate of warming 300 to 600 years prior to today for latitudes above 50° range between $0.35^\circ C$/century to $0.65^\circ C$/century is consistent with the fact that high latitude glaciers have been receding at an accelerating rate.

Thus comparison of the theoretical predictions with observed temperature change suggests that,

1) The recent rate of warming is much smaller than general accepted to date and that,

2) The prior warming trend was caused by changes in energy received from the Sun, caused by,

3) A change in the angle of tilt of Earth's axis of rotation relative to the orbital plane, a precession of the axis of rotation about a direction normal to the orbital plane, and, changes in Earth's distance from the Sun,

4) **That are all natural predictable phenomena.**

The accelerating increase in high latitude temperature and melting glaciers has been attributed by some researchers to an accelerating rate of use of fossil fuels. Fossil fuels release CO_2 and other GHG into the atmosphere. The prior belief has been that the GHG captures heat that is the cause of accelerated melting in the Arctic Regions. At first this seems like a plausible argument, but data from bore holes in Greenland show that *temperature increases BP have preceded increases in CO_2, thus refuting the assumed causal relationship between temperature rise and CO_2.* The recent Global Warming theory also cannot account for the increase in CO_2 in the past years and the corresponding decrease in temperature trends over the last 30 to 50 years. What then is the role of GHG?

The Regenerative Feedback Effect.
The net energy from the Sun absorbed by Earth depends on the emissivity of the surface. The emissivity of snow is smaller than that of land. If all the incident energy is absorbed the emissivity is 1.0. A higher average annual surface temperature is reached for a high emissivity surface than for a low emissivity surface. If at the given latitude the surface is covered with snow the emissivity of the surface is small, but for some reason an adjacent region loses snow and presents a high emissivity, heat will flow laterally from the high emissivity region[88] toward the low emissivity region. The result is a melting of the snow and an increase in the size of the high emissivity region. The boundary between the low and high emissivity region moves toward the low emissivity region. The incoming radiation appears to be more intense than before. Hence the radiation appears amplified in intensity and the rate of recession of the snow boundary increases[89]. This effect is similar to positive feedback in an electronic power amplifier called regenerative feedback.

The feedback effect is further enhanced and sustained if each successive energy input is larger than the prior input. The overall effect is an acceleration of the *rate of warming* at the given latitude. Hence, because of feedback, the calculated rates of warming at boundaries between snow cover and no snow cover will always be less than actually observed at the boundary between snow/ice pacts. Conversely, if each successive energy input is less than the prior input the *rate of cooling is accelerated* at a snow /land interface and the rate of snow coverage will increase. If the snow/ice pact interface is on water, one should not observe variability in

---

[88] Heat will flow laterally through solid land and by convection of warmer air toward the low reflectivity region.
[89] An increase in rate means melting is accelerating.

the rate of change of melting, because as the snow/ice pact melts and recedes, no emissivity changes occur at the interface, hence the rate of recession or growth remains unchanged at an ocean snow/ice pact interface.

As we can see from figures (29) though (40) the point of inflection of the summer energy received versus time curves triggers a shift from a warming to a cooling trend and glacier growth, provided the winter temperature is below 273° Kelvin.

Conversely, a shift from a cooling to a warming trend and declining or no glacier growth, results if the summer temperature increases year after year even if the winter temperature is below 273° Kelvin. There will be no glacier growth if both summer and winter temperatures are above freezing temperature.

Necessary Conditions for Glacier Growth or Decline.
According to Geologist, Earth has gone through at least two warming and cooling cycles in the past 22,000 years. These temperature swings can be attributed to celestial motion of Earth and to unpredictable catastrophic events, such as Earth/meteorite and Earth/comet collisions and volcanic eruptions. The collisions and eruptions usually cause short term intense cooling. Such an event[90]happened approximately 135 centuries BP and caused cooling lasting until about 12 centuries BP (Junger-Dryras) and another was (Krakatau 1853). The later caused very low temperatures and farmers in New England to lose a summer growing season.

 The theory does not account, nor could it account for unpredictable events but it does predict certain documented climatic events caused by celestial motion.

From geological evidence we know that ice advanced from the higher latitudes southward reaching about 41° North latitude about 95 centuries BP and then began to retreat northward. A legitimate question then is why did the ice flow start and why did it stop and why did it stop at the particular time and latitude? This question leads to yet another, namely, what are the predictable conditions for pact ice growth and decline at a particular latitude and time?

Using the observations above, the conditions for winter snow, accumulation of snow to create a glacier, and finally removal of snow causing a retreating glacier or snow pack are clear.

To cause snow to *precipitate* at a given latitude;

1) A process must exist that transports water vapor to the particular latitude, and,
2) The temperature at that latitude must be below 273° Kelvin.

To cause snow and ice to *accumulate* at a given latitude;

3) Processes that precipitate equal or increasing amounts of snow year after at a given latitude must exist, and,
4) The winter surface temperature at the given latitude must remain below or equal to 273°Kelvin, and,
5) The year after year summer temperature must not cause complete removal the prior winter snow precipitation.

To cause snow and ice to *recede* from a given latitude;

6) Processes must exist that transport snow and ice from the given latitude, and,
7) More snow and ice pact must be removed, year after year, than deposited the prior winter.
Conditions (6) and (7), are necessary for glacier retreat.

---

[90] Such an event is well document and is called the Younger-Dryas period when intense cold was cause by debris kicked up into the atmosphere by a collision with something from space.

54

A process does indeed exist during the **winter seasons** in the form of a temperature gradient that transports water vapor from the equatorial regions to the higher latitudes where it precipitates as snow. The amount of snow precipitated at the higher latitudes in winter does not significantly change year after year. Figure (42) shows the difference between energy received (or difference in temperature) in winter between 25° and 85° latitude. The gradient in temperature between low and high latitudes in a given year is related to the length of the black bars. The gradient in temperature does not change appreciably millennium after millennium. Hence, one should expect the average amount of moisture transported to the higher latitudes is not significantly changed year after year, thus meeting condition (3).

Processes of melting and sublimation provide for snow transport from a location, thus meeting condition (6).

The conditions for (2), and (4), depend on the temperature and conditions (5), and, (7) depend on the rate of change of summer temperature. The temperature and rate of change of temperature are in turn obtained from the empirical equation (see figure (37) and equation (20)) relating the energy received to temperature. Therefore, conditions for predicting the onset or retreat of glacier growth are known.

 Stated slightly differently, the <u>necessary conditions</u> for snow <u>accumulation</u> in the higher latitudes are, <u>transport of moisture</u> to the region in the winter months and <u>insufficient</u> energy <u>received in the summer</u> months to remove all of the winter snow. The later retards removal of the transported snow and the former insures input of snow. The result of *a year after year decline* in summer temperature and a winter temperature *that remains <u>below</u>* melting (273°K) and equal amounts of snow fall each winter leads ultimately to an **'Ice Age'**.

Conversely to cause an **'Ice Age Retreat'** or snow and or ice removal, the <u>necessary</u> conditions are a constant or <u>decrease in snow transport</u> to the higher latitudes in winter and an <u>increase in summer energy</u> received at the higher latitudes.

---

[93] Remember the temperature decreases as altitude increases.

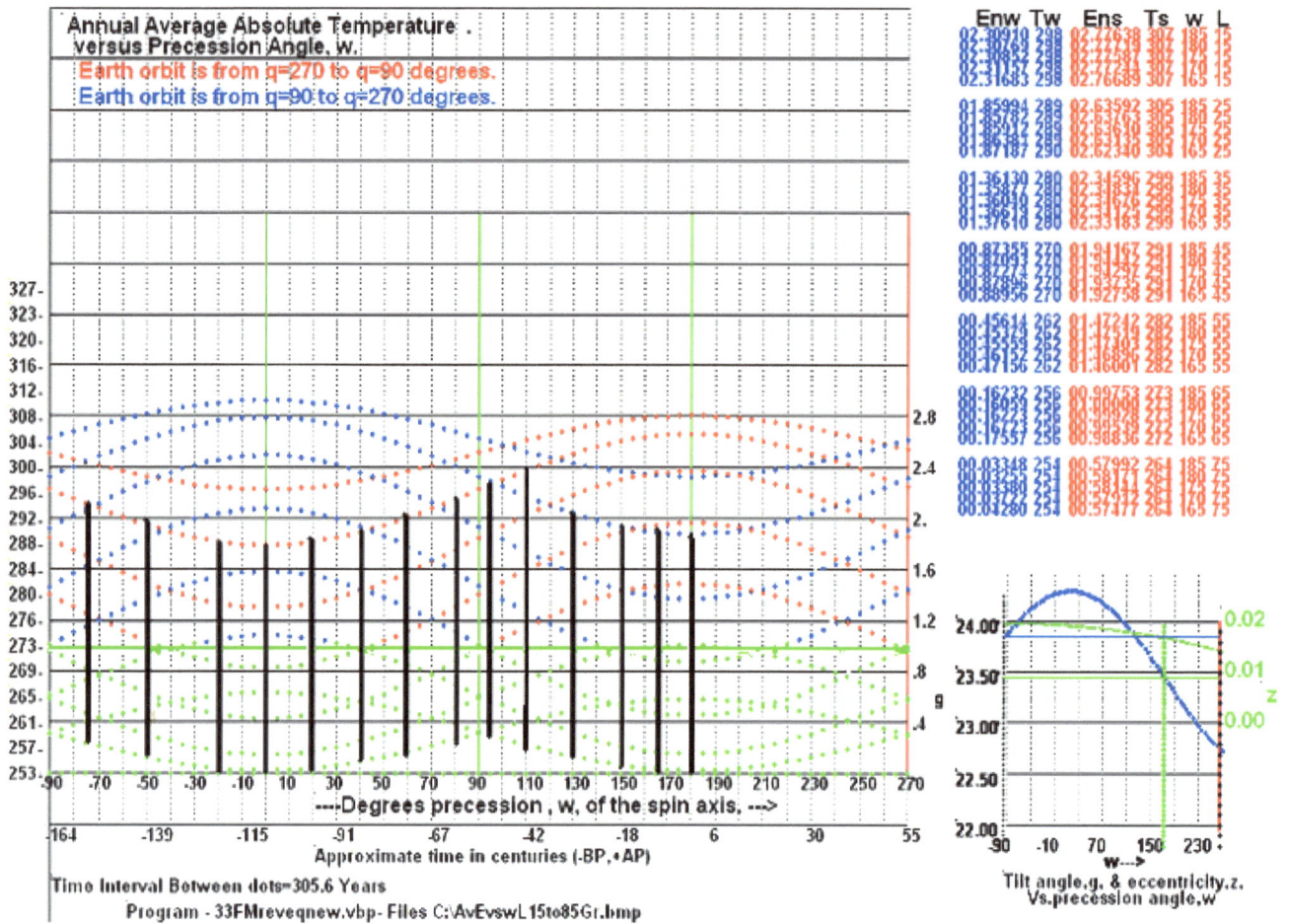

Figure (42)

The black lines are proportional to the difference in winter temperatures between the arctic and equatorial regions. The winter gradient in temperature (absent an unpredictable event) is only slightly dependent on precession angle, thus supporting the concept that the rate of mass transport of water vapor from the equatorial latitudes to the Arctic regions has changed only slightly in the past 164 centuries.

If in the cycle after snow has been transported, the temperature is above freezing and increases cycle after cycle, some of the transported snow will melt. After repeating this process for many years the snow pack thickness decreases and the line of demarcation between snow and land appears that advances toward the north. Once land is exposed the rate of absorption of energy from the Sun increases, causing more rapid melting than when the ground was covered with snow. This process is an example of how the effect of energy received is magnified and is referred to as a regenerative or positive feedback effect. Notice in figure (40) melting begins at about 95 centuries BP in the winter and that the summer temperature has started to increase and continues to increase until the present time.

How then are the conditions for glacier growth and retreat combined with the energy versus precession curves used to predict glacier melting or growth on the surface of Earth or on mountains?[93]

The following is an example of how the theory can be used to predict onset of an ice age and non-ice age time periods as a function of latitude and precession angle.

To illustrate the method we start with figure (43) which is shows energy received versus ω as the latitude goes from 40 to 76 degrees. The green horizontal line is the energy level below which the temperature is less than $273^{\circ}K$. Focus on the curves representing the energy received at north latitude 52°. Starting at $L=52^{\circ}$ and $\omega=-90^{\circ}$ and continuing to $\omega=0^{\circ}$ we see that the temperature in summer is increasing and that after passing $\omega=-75^{\circ}$ the winter temperature is always below freezing. Hence, snow transported to latitude 52° does not accumulate because each summer the temperature increases over the prior summer temperature causing more melting of prior winter snow fall. Hence, an initial snow pack at $\omega=-90^{\circ}$ recedes or is completely melted each summer season until $\omega=0^{\circ}$. After $\omega=0^{\circ}$ the summer temperature decreases each year and as long as the winter temperature is below $273^{\circ}K$, ice accumulates until $\omega=77^{\circ}$ when the temperature in winter and summer is above 273 degrees. At that point the accumulated ice and snow pack begins to melt until $\omega=110^{\circ}$. At $\omega=110^{\circ}$ the winter temperature begins to drop below 273 degrees and snow falls in winter. However, the temperature in summer begins to rise incrementally each successive year, melting more winter snow than feel in the prior year. The snow does not accumulate, but winters do become colder each successive year. Thus by 65 centuries BP ($\omega=80^{\circ}$) the ice sheet at 52° latitude begins to melt and recede and no snow falls because both summer and winter temperatures are above freezing. At $L=52^{\circ}$ and $\omega=110^{\circ}$ snow begins to fall in winter but all is melted in summer because the summer temperature increases each year thus removing all winter snow. Ultimately all snow accumulated in prior years is removed and annual snow fall followed by summer melting continues until after $\omega=180^{\circ}$. At that time, conditions for accumulation of ice and snow at 52° reappear.

Next let us focus on the conditions at latitude=64°. Note that snow and ice pact formed between $\omega=0^{\circ}$ and 180° does not meet the conditions to melt and begin to recede until $\omega=160^{\circ}$ when the summer temperature is above freezing and increasing. After $\omega=180^{\circ}$ the conditions for snow accumulation begin and glaciers begin to form.

 Figure (43) shows the time glaciers begin to retreat at latitudes 40°, 52°, and 64° degrees where we see that glaciers started to retreating at 41° latitude and are continuing to retreat today above 64° North latitude.
Judging from figure (43) the rate of glacier retreat has been increasing from the retreat that began at 41° latitude. This prediction is consistent with the increase in rate of temperature change per century as seen in figure (38). Hence the rate of glacier retreat reported in recent times is according to theory as expected.
Figure (43) shows the temperature versus precession angle for latitudes between 40 and 76 degrees. Glaciers began to retreat at $L=41^{\circ}$, and, $\omega=30^{\circ}$, at $L=52^{\circ}$, at $\omega=80^{\circ}$, and at $L=64^{\circ}$, at $\omega=160^{\circ}$. At $L=41^{\circ}$, and, $\omega=30^{\circ}$ summer temperatures started to decline year after year but remained above 273° while winter temperatures remained above 273° and increased year after year. At $L=52^{\circ}$, and, $\omega=80^{\circ}$ summer temperatures are above 273° and increase and winter temperatures decrease year after year, and at $L=64^{\circ}$, and, $\omega=160^{\circ}$, summer temperatures are above 273° and increase as winter temperatures decrease year after year. The conditions for melting are met at all latitude between $L=41$ and 64 degrees latitude. Today, between $L=52^{\circ}$ and $L=64^{\circ}$, snow falls in winter but melts prior to the next winter because of increasing summer temperatures as is predicted by the theory.

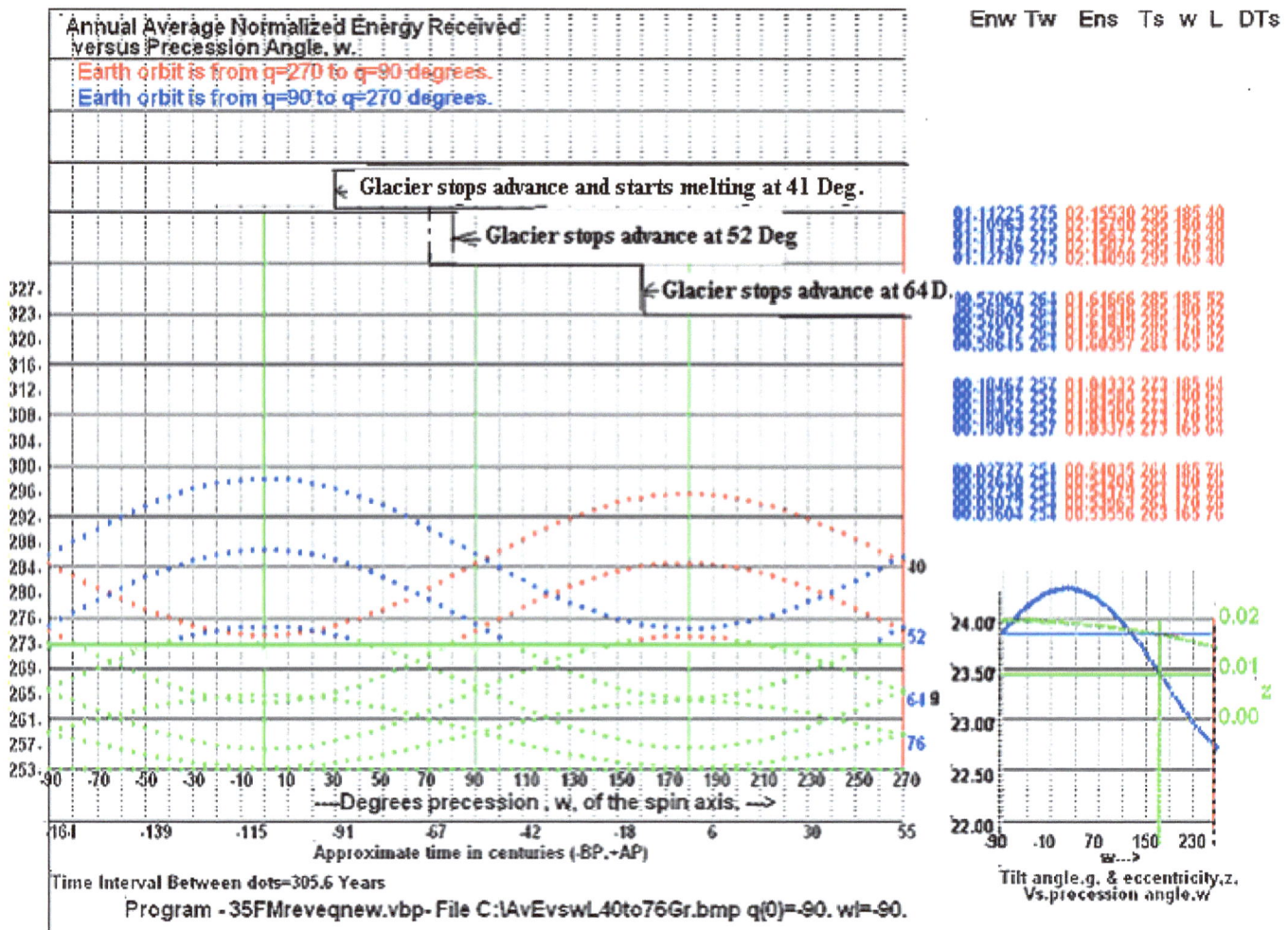

Figure (43)

Figure (43) shows the time glaciers begin to retreat at latitudes 40^0, 52^0, and 64^0 degrees where we see that glaciers started to retreating at 41^0 latitude and are continuing to retreat today above 64^0 North latitude.

Receding of Last Ice Age Predicted.

According to the best geological data to date, prior to 9,500 years BP, ice sheets extended over land into North America from the Polar Regions down to a latitude of approximately 41^0N. About 9,500 years BP (before present) the ice pacts at 41^0North latitude began to melt and recede northward. Today the ice sheets have receded to latitudes near the Arctic Circle.

The theory predicts the start of the recession of the last ice age. Referring to figure (43) we see the energy versus ω curve for the latitude of Long Island Sound which is 41^0.

Starting at $\omega=0^0$ and $L=41^0$ we see that the retreat of the last ice age is predicted. Further the theory predicts that the ice has receded to about 65 degrees North latitude as is the case today. In addition, the theory predicts onset of ice accumulation if not now but in the immediate future over land areas near 65 degrees north latitude, which has been observed recently.

Thus absent unpredictable vents, the energy received by Earth is determined by *planet Earth's motions that initiate and control the age of long term cyclic temperature changes on Earth*. Further, the temperature can change rapidly locally due to regenerative feedback processes related the emissivity of Earth's surface.

The Effect of Green House Gas Concentration on Global Temperature.

Referring to figure (37) and equation (20) we see that if no energy is received from the Sun, there is a residual constant temperature of 253° Kelvin. This is evidence that additional energy is supplied or stored by Earth or the atmosphere. We know that some of the radiation entering the atmosphere from the Sun is absorbed in the surface of Earth, raising Earth's surface temperature, and that the Earth's surface emits a spectrum of electromagnetic energy back into the air boundary layer near Earth. The wave lengths radiated by Earth are described by Plank's Law of black body radiation. In addition, we know that molecules, called Green House Gases (GHG), exist in the atmosphere that absorb selected parts of the radiation from Earth, and in doing so gain energy manifest as an increase in local air temperature. Considering all the facts above, it is reasonable to assume the residual constant temperature is the result of energy stored in the atmosphere *near the Earth's surface*. One is driven to this conclusion, in spite of the apparent fact that the concentration of a GHG such as CO_2 *follows* a temperature increase. However, if there is no mechanism to collect and store and later release energy, Earth would radiate electromagnetic energy back to space and the surface temperature would drop each night to a temperature of approximately 100° Kelvin (-173°C) , a temperature never observed before sunrise at sea level at the lower latitudes.

Consequently, the concept of energy stored in a blanket that wraps Earth, near Earth's surface, and that maintains a constant temperature independent of latitude, has evolved.

The concept of a blanket appears justified by the residual temperature shown in figure (37) because with no energy input from the Sun at night, energy is retained at the surface at night. The energy retaining ability of the blanket appears to supplements the latitude dependent energy associated with the incoming radiation from the Sun. The energy retained by the blanket is constant and appears to extend from the Equator to the Pole. Some questions arise immediately.

1) What is the mechanism that creates heat in the blanket?
2) Why does the energy released and stored in the blanket appear to be independent of
   latitude?
3) Why does the energy retained by the blanket appear to be independent of green house
    gas concentration?
4) How far does the stored energy extend into space?

A Tentative Theory of GHG Influence on Earth's Temperature.
A tentative and mostly qualitative theory, based on reasonable physical concepts follows. The theory developed is then used to gain insight as to how GHG's *may* influence Earth's temperature. In addition, the theory is used in an attempt to answer questions (1) through (4) above. The basic physical concepts used are;
1) The sun rays that ultimately reach Earth's surface release energy to the surface causing the surface temperature to rise,
 2) The surface is a black body radiator and emits a broad electromagnetic spectrum described by Plank's Law of black body radiation. The intensity of the energy radiated from the surface is a function of the temperature of the surface and the wave length of the radiated electromagnetic energy.
3) The direction of the radiation is in all directions between -90 to 90° relative to a normal to the surface.
4) The energy emitted by Earth in all directions contains wave lengths that are readily absorbed by GHG molecules causing energy of the molecules to increase[95].
5) Energy received by a GHG molecule increases the internal energy and therefore the temperature of the local atmosphere[96].

---

[95] Such molecules are referred to as Green House Gases (GHG). For example CO_2 , CH_4, and, H_2O.
[96] The internal energy of a gas is a function of the temperature only. If internal energy increases the temperature increases.

Transfer of Energy to the Atmosphere.

The following analysis of the distribution energy in the atmosphere is intended to illustrate the basic principles surrounding the blanket concept and to gain insight into the characteristics of the blanket. To simplify the basics, only the contribution of radiation normal to the surface of Earth is considered. A rigorous analysis of the energy transfer process by radiation that accounts for the fact that the black body radiation from the surface is in all directions relative to a normal to the surface was developed in 1942 by Walter M. Elsasser[97].

The energy transfer process from the surface to the GHG molecules is apparently primarily by radiation[98]. Let, I, equal the radiation from the surface of Earth in watts/ square meter and let z in meters be the distance (path length) from the surface of Earth. As the radiation from the surface enters the atmosphere the intensity decreases. The decrease in intensity depends on the product of distance traveled in the absorbing media, the number of molecules available to absorb energy from the beam, the fraction of molecules that absorb, and the local intensity of radiation. After a beam passes through a normal to a thin slab of path length thickness, dz, the decrease in intensity is,

$$dI = -kIadz, \dots\dots\dots\dots\dots\dots eq.(21)$$

where, a, is the number of GHG molecules per cubic meter (molecules/m^3). Of the absorbing molecules, k, in meters^2 per molecule (m^2/molecule)[99] is a measure of the fraction of radiation captured from the incident beam of radiation[100]. After integrating both sides of equation (21) we obtain the intensity of radiation from Earth after transmission through a path length of atmosphere of thickness, z.

$$I = [I_{surface}][Exp(-kaz)]\dots\dots\dots\dots\dots eq.(22)$$

$I_{surface}$, is the intensity in watts/m^2 of black body radiation emitted at Earth's surface. Note that the intensity of radiation declines exponentially with altitude.

Since the GHG molecules are good absorbers, they are also good emitters of radiation but not necessarily at the same wave lengths as absorbed. In other words, just like Earth's surface, the GHG molecules emit radiation. But this spectral emission is back toward the surface and in all directions including back to outer space. As a consequence, a given GHG molecule may transmit some radiation to neighboring molecules, as the neighbors can also absorb radiation from Earth. Ultimately the internal energy of the atmosphere is increased by the transfer of energy from the beams of energy emitted from the surface of Earth.

The atmosphere does not behave exactly like a black body because the GHG's do not present a continuous emission spectrum. Quite to the contrary, the emission is characterized by many sharply defined and narrow wave length bands. Never the less some of the radiation from Earth's surface is captured or in other words transferred to the GHG molecules, thereby increasing the internal energy[101] and temperature of the local atmosphere.

Clearly, the decrease in energy per unit area per unit time of the beam from Earth is equal to the energy per unit area per unit time flowing into the infinitesimally volume, dzdxdy. If we multiply dI_{GHG} by the area we obtain the energy per unit time flowing into a small volume of local atmosphere dz in height and area equal to dxdy. As a result the internal energy per unit time of this infinitesimal volume of atmosphere increases.

We first equate the change in energy/unit area per unit time from the beam, -dI, to the energy/unit area per unit time, dI_{GHG}, transferred to the GHG molecules in the atmosphere and obtain,

$$dI_{GHG} = -dI.$$

---

[97] Walter M. Elsasser, Heat Transfer by Infrared Radiation in the Atmosphere, Harvard Printing Office,1942

[98] For a detailed and basic discussion of this process see Walter M. Elsasser, Heat Transfer by Infrared Radiation in the Atmosphere, Harvard Printing Office,1942, Chapter I

[99] k defines the the ability of a particular GHG to absorb radiation.

[100] k is an intrinsic physical property of an individual GHG molecule.

[101] An increase in internal energy of a gas causes as an increase in temperature.

After taking the derivative of the radiation from the surface[102], I, and substituting for dI above we obtain,

$$dI_{GHG}=ka[I_{surface}][Exp(-kaz)]dz$$

....eq.(23)

The increment of energy flowing into an infinitesimal volume of atmosphere in a time, $\Delta\tau$, is,

$$dE=(dxdy)(dI_{GHG})(dt)=ka[I_{surface}][Exp(-kaz)]dz(dxdy)(\Delta\tau).$$

....eq.(24)

If $kaz>3$, the rate of flow of energy into the atmosphere from the surface is nil. The atmosphere above a path length at which $z>3/(ak)$ receives negligible energy input. Therefore above a certain height the atmosphere ceases to receive energy from the surface of Earth. This is consistent with the concept of a finite blanket that stores heat. To simplify the analysis we confine our attention to path lengths normal to the surface. If we sum all infinitesimal increments of energy absorbed between the surface of Earth and an altitude z and then sum over all units of time, we obtain an expression for the total energy flowing into a column of atmosphere z units high above Earth. After rewriting the cross sectional area in terms of spherical coordinates and integrating over ϕ between 0 and 2π and over L from 0 to ΔL, the total energy in a blanket of atmosphere z meters high and $R_{earth}\,\Delta L$ meters wide is,

$$E_b=I_{surface}[1-Exp(-kaz)]*(R_{earth})^2Cos(L)(\Delta L)(2\pi)(\tau)\,,$$

....eq.(25)

where, R_{earth} is the radius of Earth, L, is the latitude. The value of the variable τ depends on the time Earth receives sun light in a day at the latitude, L. Since no sun light is received at night, the blanket transfers energy back to Earth and to space. During the day the blanket receives energy from the Sun. Consequently, an average amount of energy as well an average temperature is retained by the blanket. The intensity of energy emitted from the surface, $I_{surface}$, and the time energy is received from the Sun, τ, are decreasing and increasing functions of latitude, respectively.

Equation (25) is plotted versus z in figure (44) for a range of values of ka. Here we see that total energy flowing into the blanket approaches a limit as, kaz, increases given by,

$$[E_b]_{max}=I_{surface}[1](R_{earth})^2Cos(L)(\Delta L)(2\pi)(\tau).$$

....eq.(26)

As should be expected the energy stored in the blanket cannot exceed the source of energy. In other words the blanket cannot store more energy than is supplied by the source. Further if $kaz>3$ the energy stored in the blanket is independent of the product of the term, kaz, hence independent of the concentration of GHG. The energy stored, after saturation, is a function only of the energy emitted from the surface of Earth. In addition, if the quantity, kaz, is greater than ~3, all the surface emitted energy available is stored. Therefore, further increase in GHG concentration does not increase the temperature.

After reaching saturation, the temperature is constant and independent of an increase in concentration of GHG molecules.

Measurements[103] show that temperatures at northern latitudes recorded between 30 to 50 years BP (see Figure(41)) are essentially constant temperature throughout a time period in which GHG concentrations increased at least 4 fold[104]. An explanation for this observation is found in the tentative theory. According to the theory, the GHG concentration has reached and exceeded the saturation level thus a limit has been reached in the influence of GHG concentration on temperature.

This conclusion considers only energy radiated normal to the surface. However if one applies the same reasoning to a bean of energy radiating at any angle between -90 to +90 from the normal, one arrives at a

---

[102] See equation (22)

[103] See figures (41), page 52. also see footnote (69)

[104] A.B. Robinson, N.E. Robinson, and , Willie Soon Journal of American .Physicians and Surgeons (2007) 12,79-90 Figure (2)

result similar to that obtained for the normal beam only. Allowing for absorption of radiation for all beams from the surface in effect increases the absorption sites. Thus one should expect the depth of penetration of the emitted radiation to decrease and the GHG concentration beyond which one observes no further increase in temperature to decrease.

The Distance the Blanket Extends From the Surface.
We see from equation (24) and figure (44) that the local rate of energy flow approaches zero if the height of the column, z_{max} exceeds approximately 3/(ka). The thickness of the blanket is controlled by the reciprocal of the product of the number of molecules/unit volume and the quantity k that accounts for the absorption properties of the GHG molecule. Increasing the concentration of GHG molecules does not increase the energy stored, it merely decreases the height of the blanket.

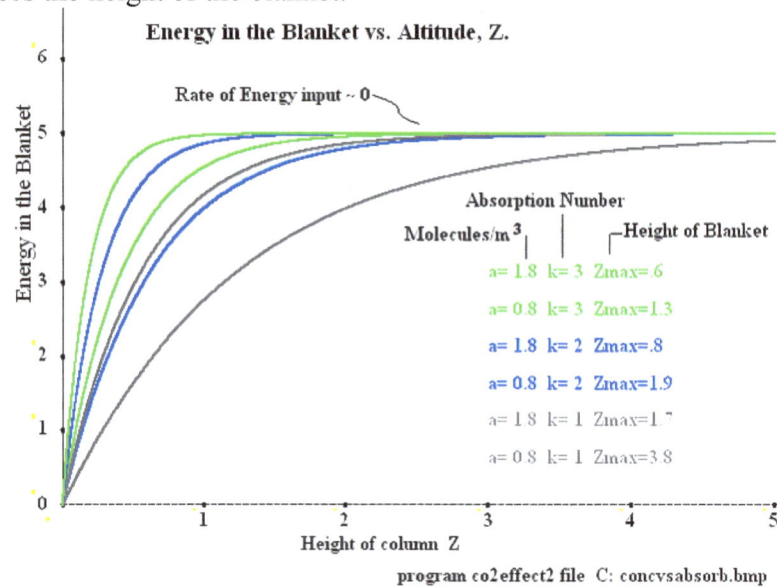

Figure (44)

Equation (25) is plotted in figure (44) and illustrates <u>schematically</u>, how energy may be stored in Green House Gases. The magnitude of energy stored after saturation, is independent of product of concentration, a, and, absorption number, k, of a GHG near Earth. The energy stored in the atmosphere saturates if z>3/(ka). After saturation, <u>no additional energy is stored</u> thus defining the height of the storage blanket. If z>3/(ka), increasing, a, only decreases the height of the blanket. Increasing, k, which means adding gases that trap radiation more efficiently only decreases the height of the blanket. Hence, after saturation is reached all GHG's, regardless of their concentration or absorptive characteristics, the amount of energy stored in the atmosphere remains unchanged. The amount of energy stored is a function of the intensity of the black body radiation from the surface of Earth and after saturation is independent of the absorption characteristics and amount of the particular GHG.

Dependence of Blanket Temperature on Green House Gas Concentration.
As the number of molecules per meter^3 increases (for example an increase in CO_2 concentration), the energy transferred into the column of atmosphere near the surface approaches a constant. Thus, increase of GHG beyond the level of saturation, contributes <u>nothing</u> to the amount of energy stored in the blanket. The heating of the blanket depends solely on energy emitted by the surface of Earth.

Stated differently, *after saturation increased GHG concentration in the atmosphere does not increase the temperature of the blanket.* The temperature of the blanket depends solely on the energy radiated from the surface of Earth and that in turn depends on the energy received from the Sun. Prior to reaching saturation, the temperature depends on concentration.

A Blanket Temperature Independent of Latitude.
After saturation, the energy, $[E_b]_{max}$, is a function of the intensity of the black body radiation emitted by Earth. The black body radiation is in turn a function of latitude that decreases with latitude. Hence, energy should decrease as latitude increases, leading to the conclusion that the energy stored in the blanket should decrease as the latitude increases, but there are compensating processes operative.

62

Referring to equation (26),

$$[E_b]_{max} = I_{surface}[1](R_{earth})^2 Cos(L)(\Delta L)(2\pi)(\tau),$$

we see that the energy transferred to the blanket is a product of the radiation from the surface, the time radiation is received and the latitude. The radiation from the surface and time it is received are decreasing and increasing functions of latitude, and therefore tend to offset their respective effects on energy stored. Because the equation for energy transferred into the blanket uses only the normal component of surface radiation, the term, $I_{surface}$, declines with increasing latitude more rapidly than one would expect. A more rigorous analysis that includes energy emitted from -90 to+90 degrees from the normal would show a smaller decline in energy transferred to the blanket with latitude. In addition, a temperature gradient exist that drives energetic GHG molecules at lower latitudes toward the higher latitudes thereby increasing the concentration of GHG molecules at higher altitudes and increasing product, ka, and therefore the energy stored.
Measurements of CO_2 concentration versus latitude show an increase in CO_2 concentration of approximately 6% from the equator to 75° N. latitude[105].

To simplify the concept of transfer of energy by radiation, we have assumed radiation is only emitted normal to the surface of Earth. In fact, radiation comes from all directions and offsets a decrease in the magnitude of surface radiation at higher latitudes. Hence, it is not unreasonable to expect the energy to be approximately constant or decline slightly as one moves along a meridian toward the Pole. Measurements of temperature (an indicator of energy stored) at altitudes less than 8 km decline of the order of only 0.1°C between 30° to 75° N. latitude[106]. This data supports the result shown in figure (37) and the concept of a blanket that stores a constant amount of heat independent of latitude.

Therefore considering all the processes operating one should expect that the energy stored in the blanket at higher latitudes should be higher than expected because of *energy transfer by radiation* toward the higher latitudes and the increase in concentration of radiation absorbing molecules, all factors supporting the blanket concept.

Summary and Conclusions
Global warming reported prior to today can be accounted for by changes in energy received from the Sun caused by naturally occurring changes in the Earth's precession and tilt angles and minor changes in Earths orbital eccentricity. The results of the calculations make sense physically and appear to agree with observations. The analysis is based solely on the motion of Earth's tilt and precession angles and changes in orbital eccentricity, that cause changes in energy received by Earth that are largely beyond our control.

Only established knowledge of the motions of Earth and the effect these motions have on energy received by Earth have been used in developing the theory. In addition to predicting the increases observed in temperature per century, the theory correctly predicts the retreat of the last ice age and the continuing recession of the last ice age. The most salient prediction however is that the warming cycle has or will reverse and a cooling cycle will begin or has begun that will last for at least 5,000 years after present.

An analysis of green house gases, based partly on theory and data obtained by direct measurement, suggests that Green House Gases store energy in near Earth's surface caused by Earth re-radiating energy to GHG absorbers near the surface of Earth. The average temperature of the blanket appears to be nearly independent of latitude because heat is transferred to the blanket primarily by radiation in all directions from Earth's surface and energy emitting GHG molecules. Horizontal and vertical temperature gradients appear to mix

---

[105] Ibid., S.Fred Singer ,Fig.(21) page 21
[106] Ibid., ., S.Fred Singer ,Fig.(8) page 7

high and low latitude radiated energy very efficiently, thus maintaining the nearly the same blanket temperature at all latitudes.

The temperature of the blanket is determined by intensity of the radiation from the surface. Theoretical analysis suggests that the blanket temperature, once a saturation concentration is reached, <u>does not continue to increase as green house gas concentration increases</u>.

Since, theory predicts that the energy received from the Sun is currently increasing or at a maximum, and GHG concentration is also increasing, one might claim the cause for global warming is the steady increase in GHG concentration. However, temperatures changes caused by changes in energy received from the Sun appear to be in agreement with temperature increase per century data. Further, the theory suggests that the heating effect of increasing GHG concentration approaches a limit asymptotically. Measurements of temperature with properly sited and specified instrumentation show that the temperature at least 30 years BP has been essentially constant as world hydrocarbon use, a source of GHG, has increased about 3 to 4 fold in the last 100 years. Hence, the influence of GHG on temperature appears to have reached a limit caused by complete absorption of surface radiation by GHG. Therefore, the variables controlling Earth's temperature are eccentricity, tilt angle and precession angle, all parameters that are beyond our control with today's technology.

Using recent global temperature data, the theory developed above, and analysis, a logical conclusion is,

1) *Global warming and cooling is caused mainly by changes in energy received from the Sun caused in turn by changes in the direction of the spin axis and eccentricity of Earth's orbit and that,*
2) *Earth has been warming since the last North American glacier began to recede 95 centuries BP and that,*
3) *The warming cycle is currently at or slightly past the maximum and that,*
4) *Energy received from the Sun is in or beginning a decline, going forward, causing global cooling and that,*
5) *Green House Gases create a constant temperature blanket of air near the surface of Earth whose temperature slightly, if at all, dependent on GHG concentration and that,*
6) *Consequently increasing GHG concentration is not the cause of global warming*

Thus steps to reduce global warming by eliminating green house gas may not produce the expected result and might accelerate and intensify the predicted cooling trend of a new ice age, which could be dangerously counter productive, indeed.

George T. Croft, Ph.D.   October 22, 2009
22 Coventry Court, Bluffton, SC 29910

Program used to Calculate E $_{band}$ (ε, γ, θ, ω, L).

Because the computer program used does not recognize the Greek alphabet, the symbols used in the text have been replaced by corresponding English characters. Hence θ becomes q, ε becomes e, γ becomes g, and etc. To simplify and shorten the computational time of the program, a 364 day year is used. Using a 364 day year, the time scale is shortened 0.34% per year or by 75.5 years after 22,000 years. The orbital position, θ, of Earth is divided into 360 /364 increments.

The orbital positions of Earth are indexed by the integer, m. Each year θ is increased by $2\pi/364$ radians, thus, q(m+1)= q(m)+ $2\pi/364$. If m=0, θ=-90°, and if m=1, θ=-89.0011°. If m=90, θ=0°, etc (see figure (28a)).

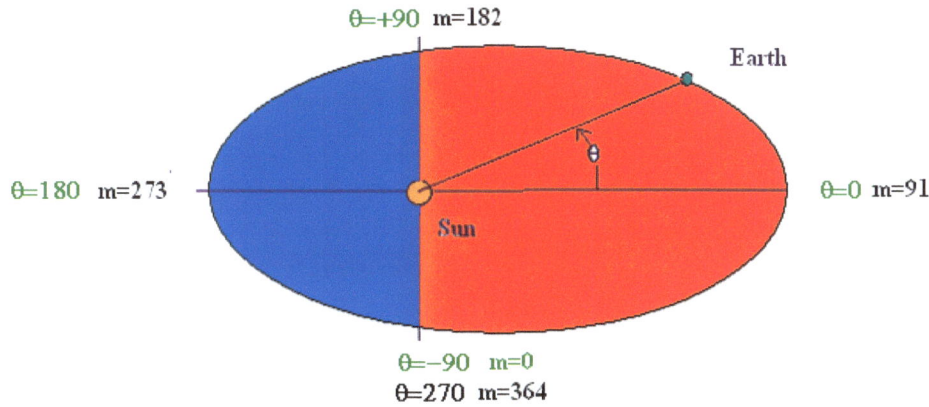

θ=+90 m=182

Earth

θ=180 m=273

θ=0 m=91

Sun

θ=-90 m=0
θ=270 m=364

q(m+1)=q(m)+pi/182   After h steps q(h)=-91/182*pi+h*1/182
q(0)=-91/182*pi   let m=0   q(0)=-91/182*pi--->-90
      let m=91   q(91)=-91/182*pi+91/182*pi-->0
      let m=182   q(182)=-91/182*pi+182/182*pi-->+90
      let m=273   q(273)=-91/182*pi+273/182*pi-->+180
      let m=364   q(364)=-91/182*pi+364/182*pi=--->+270

Figure (28a)

To obtain the energy received as Earth travels from 270° to 90° the trajectory (red) of Earth m is indexed from m=0 to m=181. To obtain the energy received as Earth travels from 90° to 270° the trajectory (blue) of Earth is indexed from m=182 to m=364. If m=100 θ=8.90°

The Program

The program is written in Visual Basic 6
Private Sub Command1_Click():
Name of Program - tempvstime34FMrev.vbp- Files C:\Av_En_vs_w_LREV.bmp
Rem this program integrates the energy received by Earth as a function of latitude and precession
Rem the Energy received in a given year is the total received for that year.
Rem The energy is calculated assuming that the steady state is reached for each increment in time in a day and for each increment in a year
Rem L=latitude, f=longitude, w= precession, and, g=tilt angle
Private Sub Command1_Click():
Rem Name of Program - tempvstime33FMreveqnew.vbp- Files C:\Av_En_vs_w_LREV.bmp
Rem this program integrates the energy received by Earth as a function of latitude and precession
Rem the Energy received in a given year is the TOTAL received for that year.
Rem The energy is calculated assuming that the steady state is reached for each increment in time in a day and for each increment in a year
Rem L=latitude, f=longitude, w= precession angle relative to current summer solstice position of Earth q=0, and, g=tilt angle
Dim E#(1010, 1010): Dim f#(1010): Dim Ez#(1010, 1010): Dim q#(1010): Dim SSSTemp(1000, 1000): Dim SSTemp(1000, 1000)

Dim h#(1010): Dim Temp(1010, 1010): Dim Ezz#(1010, 1010): Dim STemp(1010, 1010): Dim TempS(1010, 1010):
Dim SEzz#(1000, 1000): Dim SSEzz#(1000, 1000): Dim SSSEzz#(1000, 1000):
Dim TE#(1010, 1010): Dim TEz#(1010, 1010): Dim TEzz#(1010, 1010):
Rem setup screen parameters
BackColor = QBColor(15)
Rem The screen horizontal width is 3.1*pi and the vertical axis is +1 to -1 units
cls: pi = 4 * Atn(1): sc = 1: Scale (-0.1 * pi, 1 * sc)-(3 * pi, -1 * sc): Top = 0: Left = 0:
Height = 12500: Width = 1.3 * Height: C = 131400: AutoRedraw = true
Rem introduced for later calculations
Rem define physical parameters most not used in program but are here for reference
Rem density ice=.917*10^3,air=1.29,water=1*10^3 Kg/m^3
Rem specific heat ice=2090, air=, water=4168 J/kg
Rem define vatiables
Rem z= eccentricty of orbit
 Rem g=tilt from normal to orbital plane
 Rem w= precession angle
  Rem L= Latitude
 Rem tilt angle gg = -23.45
   Rem Cp = 2090
Rem rho = 917: Rem h=~2(Dt)^1/2, t=h^2/(4*D)
 Rem a=major axis for simulation
 a = 0.1:
 Rem zo=reference eccentricity, z=eccentricity
 zo = 0.017:
 Rem emmisivity and Boltzmann constant
 em = 0.9: sig = 5.67 * 10 ^ -8
   Rem net power received from sun at surface of Earth in watts/meter^2
 Eoo = 1350
 Rem **
 Rem set magnitude of increments in f and q.
 Rem Longitude, f , steps 2pi/360 or 1 degree/step until 360 degrees rotation about spin axis. Thus increment in f is:
 df = 2 * pi / 360:
Rem q steps around Sun in 365.25 days. Earth goes around spin axis once in 24 hours.
Rem but to simplify calculations I have let a year=364 days. This is done to allow selection of sections of an orbit.
Rem q= angular displacement of earth about the Sum where q=0 is on the horizontal axis and earth sun distance is greatest.
dq = 2 * pi / 364
Rem thus Earth steps (2pi)/366 radians/day or 2*pi/366/pi*180=0.9832 degrees each day of orbit.
Rem Since the adjusted year is longer so must the day be longer. 24/365.25=x/366 therefore x=366/365.25*24=24.0493 hours or 3 minutes longer
Rem The adjusted year is 18 hours longer
Rem stp is the increment in the precession angle=5 degrees
stp = 5 / 180 * pi
Rem adjust scale factor
adj = 0.25
Rem the time for Earth to rotate dq=2*pi/360 units is 2*pi/360*24/2*pi=24/360=4 minutes Hence energy is calculated every 4 minutes.
 Rem begin plotting the the coordinates
 Rem lay in vertical and horizontal ordinates
 Rem lays in a set of lines spaced 0.1 units
 DrawWidth = 1
 Rem define an adjustment factor for locating printed numbers on the scale
 shf = -0.5
 For m = -0.6 * 1 To 1 Step 0.1: For n = -0 To 2 * pi Step 0.001 * pi / 2: PSet (n, m): PSet (0, n): Next n: Next m
  For m = -0.6 * 1 To -0.2 Step 0.1: For n = 2.23 * pi To 2.72 * pi Step 0.001 * pi / 2: PSet (n, m): PSet (0, n): Next n: Next m
 Rem lays in a set of lines spaced horizontally by 10 units of angle but written vertically by 0.01 units
 For m = -0.6 To 1 Step 0.01: For n = -0 To 2.05 * pi Step 10 / 180 * pi: PSet (n - 0# * pi, m): Next n: Next m
 For m = -0.6 To -0.1 Step 0.01: For n = 2.23 * pi To 2.72 * pi Step 10 / 180 * pi: PSet (n - 0# * pi, m): Next n: Next m
 FontSize = 12
 Rem label the horizontal coordinate
 PSet (100 / 180 * pi + pi / 16, -0.039 + shf): Print "----Degrees precession , w, of the spin axis. --->":
 Rem place labels for En,w,L
 FontSize = 12: ForeColor = QBColor(0): PSet (385 / 180 * pi, 1.005), QBColor(15): Print ; " Enw ": PSet (408 / 180 * pi, 1.005),
QBColor(15): Print " Tw": PSet (430 / 180 * pi, 1.005): Print " Ens ": PSet (458 / 180 * pi, 1.005): Print "Ts": PSet (475 / 180 * pi, 1.005):
Print "w": PSet (488 / 180 * pi, 1.005): Print "L": PSet (500 / 180 * pi, 1.005): Print "DTs"
 Rem print longitude on the horizontal scale in degrees
 FontSize = 10: For n = -0 To 2 * pi Step pi / 4.5 / 2: PSet (n - 0.023 * pi, shf): Print Format(n / pi * 180 - 90, "###"): Next n:
 Rem time
 FontSize = 10: For n = -0 To 2 * pi Step pi / 4.5: PSet (n - 0.01 * pi, shf - 0.1): Print Format((n / pi * 180 - 270) * 605.6 / 10 / 100, "###"):
Next n:

PSet (100 / 180 * pi + pi / 16, -0.14 + shf): Print "Approximate time in centuries (-BP,+AP)":
Rem begin printing of system parameters
Rem PSet (5 / 180 * pi + pi / 16, -0.65): FontSize = 12: Print " Tilt angle="; gg0; " deg. at w="; 0 / pi * 180; " deg. and "; Format(gg360,
"##.#"); " deg.at w="; wu / pi * 180; " deg"   PSet (1 / 180 * pi, -0.7): Print "Time Interval Between dots="; Format(stp / pi * 180 / 360 *
22000, "###.#"); " Years"
  Rem PSet (175 / 180 * pi, -0.7): Print "Eccen.="; z; " at w=360 degrees"
  PSet (32 / 180 * pi, -0.75): Print "Program - 33FMreveqnew.vbp- Files C:\AvDTEvswL20to80G.bmp"
  PSet (400 / 180 * pi, -0.69): Print "Tilt angle,g, & eccentricity,z,"
  PSet (410 / 180 * pi, -0.72): Print "Vs.precession angle,w"
Rem print the normalized component of the suns energy/unit area on the vertical scale in middle of page
    ForeColor = QBColor(12)
For m = shf To 0.45 * 0.6 Step 0.1: PSet (2 * pi, 1 * m + 0.02): Print Format((m - shf) / adj, " #.##"): Next m:
  Rem write absolute temperature on vertical scale
    ForeColor = QBColor(0)
For m = shf To 0.45 Step 1 / 20: PSet (-0.093 * pi, 1 * m + 0.02), QBColor(14): Print Format(((m - shf) / adj * 19.58 + 253), "###"); "-":
Next m
    DrawWidth = 2: For nn = 0 To 2 * pi Step pi / 3000: PSet (nn, -0.28 + 0.02 * 1.02), QBColor(10): Next nn:
    DrawWidth = 1
Rem lay in a line showing current location of spin axis as it precesses about the pole in green and in red lay in completion of one
precession
  Line (0, -1)-(0, 1)
    Line (360 / 180 * pi, -0 + shf)-(360 / 180 * pi, 0.6), QBColor(12)
    Line (270 / 180 * pi, -0 + shf)-(270 / 180 * pi, 0.6), QBColor(10):
    Line (90 / 180 * pi, -0 + shf)-(90 / 180 * pi, 0.6), QBColor(10):
    Line (180 / 180 * pi, -0 + shf)-(180 / 180 * pi, 0.6), QBColor(10):
    FontSize = 12
Rem q= angle Sun rays make with respect to Sun system
  Rem the index for sun is m. Index for longitude is n.
  ForeColor = QBColor(12): PSet (10 / 180 * pi, 0.9): Print " Earth orbit is from q=270 to q=90 degrees."
 ForeColor = QBColor(9): PSet (10 / 180 * pi, 0.85): Print " Earth orbit is from q=90 to q=270 degrees.": ForeColor = QBColor(0)
    Rem label eccentricity curve
  ForeColor = QBColor(10): PSet (500 / 180 * pi + pi / 25, -0.32), QBColor(0): Print " z"
  PSet (490 / 180 * pi, -0.171), QBColor(0): Print " 0.02"
  PSet (490 / 180 * pi, -0.271): Print " 0.01"
  PSet (490 / 180 * pi, -0.371): Print " 0.00"
  Rem label plot of E vs.w
  ForeColor = QBColor(0): PSet (5 / 180 * pi, 0.99): Print " Annual Average Absolute Temperature ."
 ForeColor = QBColor(0): PSet (10 / 180 * pi, 0.95): Print "versus Precession Angle, w. "
            Rem SET LATITUDE RANGE AND INCREMENT
                Rem L = latitude set latitude range
    For l = 20 / 180 * pi To 80 / 180 * pi Step 10 / 180 * pi:
  stp = 5 / 180 * pi:
  Rem v is a index allowing calculation of time
  v = v + 1
  p = 0
      Rem START RANGE AND INCREMENT OF PRECESSION ANGLE
  Rem the wa used here in place of w
  wl = -90 / 180 * pi: wu = 270 / 180 * pi + 0 / 180 * pi
    For wa = wl To wu Step stp:
Rem PSet (0, U / 30): rem Print "U="; U; "   q="; q(m) / pi * 180 - 360
  Rem gg= tilt angle in degrees. Equation below is a curve fit to Imbire and Imbrie.
    Rem calculate eccentricity equation below is a curve fit to Imbire and Imbrie.
        Rem new equation for gamma vs omega
            If wa < pi Then GoTo 44 Else GoTo 54
44 gg = 23.45 - 0.85 * Sin(22 / 40 * (wa - pi)):
GoTo 88
54 If wa > pi Then GoTo 66
66 gg = 23.45 - 0.7 * Sin(22 / 32 * (wa - pi))
88   g = -gg / 180 * pi
        Rem print ecentricity vs omega
  Rem the current omega is +180 degrees according to convention adopted in text.
  Rem v=1 if wa=pi we have current conditions. Please check for the convention in figure 1
90 vvv = 1
  If wa >= vvv * pi Then GoTo 100 Else GoTo 200
100 z = 0.017 - 0.004 / pi * (wa - vvv * pi) - (0.003 / (2 * pi ^ 2)) * (wa - vvv * pi) ^ 2: GoTo 300
200 z = 0.017 - 0.004 / pi * (wa - vvv * pi) - (0.0055 / (4 * pi ^ 2)) * (wa - vvv * pi) ^ 2:
300 Rem PSet (wa, -gg / 23.45), QBColor(12):
GoTo 302

If wa > -40 / 180 * pi - dq And wa < -40 / 180 * pi + dq Then PSet (wa + 90 / 180 * pi, 0.46): Print " wa="; Format(wa / pi * 180, "0##.#"); " g="; Format(gg, "##.#"); " z="; Format(z, "#.####") Else

If wa > 0 - dq And wa < 0 + dq Then PSet (wa + 90 / 180 * pi, 0.5): Print " wa="; Format(wa / pi * 180, "0###.#"); " g="; Format(gg, "##.#"); " z="; Format(z, "#.####") Else

If wa > 90 / 180 * pi - dq And wa < 90 / 180 * pi + dq Then PSet (wa + 90 / 180 * pi, 0.5): Print " wa="; Format(wa / pi * 180, "0##.#"); " g="; Format(gg, "##.#"); " z="; Format(z, "#.####") Else

If wa > 180 / 180 * pi - dq And wa < 180 / 180 * pi + dq Then PSet (wa + 90 / 180 * pi, 0.46): Print " wa="; Format(wa / pi * 180, "0##.#"); " g="; Format(gg, "##.#"); " z="; Format(z, "#.####") Else

302 Rem old formula z = 0.017 - 0.00005 * (wa / pi * 180 - 270) - 0.000000154 * (wa / pi * 180 - 270) ^ 2:
          g = -gg / 180 * pi: DrawWidth = 2
                If test < 1 Then GoTo 45 Else GoTo 800
    Rem plot tilt angle vs. wa but first print in coordinates
       Rem vertical base line
45   For n = -0.45 To 0.05 Step 0.01: PSet (400 / 180 * pi, n - 0.19), QBColor(0): Next n: FontSize = 10
      Rem horizontal coordinate line
      DrawWidth = 3: For n = -0.4 To -0.2 Step 0.1: PSet (400 / 180 * pi - pi / 75, n), QBColor(0): Next n: DrawWidth = 2
      Rem label vertical coordinate
      PSet (367 / 180 * pi + pi / 50, -0.2 + 0.08 - 0.21), QBColor(15): Print "g"
      Rem write vertical scale for tilt angle numbers
      For n = -0.4 To -0# Step 0.1: PSet (380 / 180 * pi, 1 * n + 0.02 - 0.21), QBColor(0): Print Format(4 + n / 0.2 + 20, "##.#0"): Next n:
Rem write horizontal numbers for w from 0 to 360 in steps of 40
ForeColor = QBColor(0):
For n = 400 / 180 * pi To 490 / 180 * pi Step stp * 4 * v: PSet (n - n * pi / 1000 - stp, -0.62), QBColor(15):
Print (Format(4 * n / pi * 180 - 1290, "###") - 400): Next n:
      Rem write label for horizontal scale
      ForeColor = QBColor(0): PSet (440 / 180 * pi, -0.65): Print " w--->"
      Rem draw line thru w=360 and make it red
      Rem For n = -0.41 To -0# Step 0.001: PSet (490 / 180 * pi, n - 0.2), QBColor(12): Next n
       Line (490 / 180 * pi, -0.41 - 0.2)-(490 / 180 * pi, -0.2), QBColor(12)
      Rem print dots every 0.2 units over the red
       DrawWidth = 3: For n = -0.42 To 0.02 Step 0.02: PSet (490 / 180 * pi, n - 0.22), QBColor(0): Next n
        Rem draw line thru w=180 and make it green
        For n = -0.44 To -0# Step 0.01: PSet (467 / 180 * pi, n - 0.2), QBColor(10): Next n
        Rem plot tilt angle from 22 to 25 degrees vs w from 0 to 360
          PSet (400 / 180 * pi + stp * v / 4 + 0 / 180 * pi, ((gg / 23.45 - 1) * 5 - 0.15 + 0.039 - 0.21)), QBColor(9):
      Rem PSet (400 / 180 * pi + stp * v / 4, ((22 / 23.45 - 1) * 0 - 0.15 + 0.039 - 0.2)), QBColor(0): DrawWidth = 2
      Rem For n = 400 / 180 * pi To 590 / 180 * pi Step stp * v / 2: PSet (n, -0.4 - 0.2), QBColor(3): Next n
      DrawWidth = 1: Line (400 / 180 * pi, -0.15 + 0.039 - 0.2)-(490 / 180 * pi, -0.15 + 0.039 - 0.2), QBColor(10)
      Rem plot ecentricity vs wa
      DrawWidth = 2: PSet (400 / 180 * pi + stp * v / 4, ((z - 0.017) * 10 - 0.24 + 0.007)), QBColor(10):
      Rem draw current eccentricity in green
      DrawWidth = 1
      Rem PSet (400 / 180 * pi + stp * v / 4, ((0.017 - 0.017) * 10 - 0.24)), QBColor(10):
      Line (400 / 180 * pi, -0.24 + 0.01)-(490 / 180 * pi, -0.24 + 0.01), QBColor(9)
        Rem BEGIN CALCULATION OF ENERGY RECEIVED AS A FUNCTION OF Z,G,Q,W,L
Rem step earth about the sun in increments of dq
800 For m = 0 To 364 Step 1
        q(0) = (-91) / 182 * pi
        q(m + 1) = q(m) + dq:
      Rem step earth about the the spin axis in increments of df=2pi/360
For n = 0 To 359 Step 1
        f(0) = 0
        f(n + 1) = f(n) + df:
        Rem If wa + stp > 180 And wa - stp < 180 Then Print " wa="; Format(wa / pi * 180, "###.#"); " q="; Format(q(m) / pi * 180, "###.#") Else
       Rem PN from coordinate systen reconsidere22 Equation(7)
    Rem PN=E=((Cos(L)*Cos(f)*Cos(g)*Cos(w)-Cos(L)*Sin(f)*Sin(w)+
    rem Sin(L)*Sin(g)*Cos(w))*Cos(q)+(Cos(L)*Cos(f)*Sin(w)*Cos(g)+ Cos(L)*Sin(f)*Cos(w)+ Sin(L)* rem Sin(w)*Sin(g))*Sin(q))
    E(n + 1, m + 1) = ((Cos(l) * Cos(f(n)) * Cos(g) * Cos(wa) - Cos(l) * Sin(f(n)) * Sin(wa) + Sin(l) * Sin(g) * Cos(wa)) * Cos(q(m)) + (Cos(l) * Cos(f(n)) * Sin(wa) * Cos(g) + Cos(l) * Sin(f(n)) * Cos(wa) + Sin(l) * Sin(wa) * Sin(g)) * Sin(q(m)))
    DrawWidth = 1
    Rem Correct for orbit distance and Earth angular position
    Rem PNR = (1 - z * Cos(q(m))) ^ 2 / (1 - z ^ 2) ^ 2 * PN
    Rem E is from eq (7) Monograph/coordinate system reconsider2 modified to allow program requirement to step digitally
    Rem Ez =PNR is from eq (12) Monograph/coordinate system reconsider2 corrected for eccentric orbit
    Ez(n + 2, m + 2) = (1 - z * Cos(q(m))) ^ 2 / (1 - z ^ 2) ^ 2 * E(n + 1, m + 1)
    Rem Ez(n + 2, m + 2) = ((1 - z * Cos(q(m))) / (1 - z ^ 2)) ^ 2 * E(n + 1, m + 1):
    Rem if E is negative then no energy is received so it is set equal to 0 otherwise negative energy is summed

If Ez(n + 2, m + 2) < 0# Or Ez(n + 2, m + 2) = 0# Then GoTo 902 Else GoTo 922
902 Let flag = 0#: GoTo 932
922 flag = 1#
932 Ezz#(n + 3, m + 3) = flag * Ez(n + 2, m + 2):
 Rem Code below inserted to establish phase relationship between precession angle and orbital position of Earth------
  Rem If m = 0 And wa > 180 / 180 * pi - dq And wa < 180 / 180 * pi + dq Then GoTo 940 Else GoTo 5000
Rem m=0 corresponds to q=90 degrees and if wa=270 degrees wa looks like \|/and a day is longer than 12 hours
Rem if m=0 q=90 degrees and if wa=90 a day is 12 hours
Rem m=90 corresponds to q=90+90=180 degrees and if wa=270 degrees wa la day is exactly 12 hours
940 Rem PSet (f(n) + stp, Ezz#(n + 3, m + 3)), QBColor(m / 30): PSet (wa, 0): Print "q="; (q(m) / pi * 180); " wa="; Format(wa / pi * 180,
"###.#"):
Rem Start summing Ezz#(n+3,m+3) *Cos(L) over n=0 to 359 for one revolution about the spin axis for each orbital position of Earth
5000 SEzz#(n + 4, m + 3) = SEzz#(n + 3, m + 3) + Ezz#(n + 3, m + 3) * Cos(l):
502   Rem SSEzz#(n + 5, m + 3) = SSEzz#(n + 4, m + 3) + SEzz#(n + 4, m + 3):stop
If n = 359 Then GoTo 501 Else GoTo 504
Rem SSSEz#(n+5,m+4) sums energy for each day of an ORBIT OF EARTH for a given w, and L
501 SSEzz#(n + 4, m + 4) = SSEzz#(n + 4, m + 3) + SEzz#(n + 4, m + 3):
504 Rem after summing all revolutions about the spin axis and all positions of Earth in orbit , capture the total energy received for m from
0 to 180
Rem q(0) =91/180*pi and q(m+1)=q(m)+dq thus when m=0 q=pi/2 or 90 degrees and when m=182 q(90+m)=q(270)
Rem Thus if m=180 and n=359 the energy received between q=90 and q=270 or -90 and +90 is captured
582   If m = 182 Then GoTo 542 Else GoTo 72
542 Rem if m=182 then earth has traveled from m=-91 to +91 and
Rem if n=359 all energies received for 360 degree rotations of earth from q=-90 to +90 have been summed
Rem this allows color change to green if the temperature is less than 273K
If (SSEzz#(n + 4, m + 4) / 7368.60821016331) < 1.0214 Then cg = 10 Else cg = 12
 DrawWidth = 3: D = 65520: PSet (stp * (v - 1), shf + (2.25 / D) * SSEzz#(n + 4, m + 4)), QBColor(cg)
 700
 Rem now restart the rotation about the spin axis.
 GoTo 722
72   If m < 182 Then GoTo 722 Else GoTo 804
Rem Begin calculating total energy received between q=270 and q=90 or between +90 and -90
804 Rem q(180) = 270 / 180 * pi or -90 and +90
     Rem f(0) = 0:
q(m + 1) = q(m) + dq
f(n + 1) = f(n) + df:
Rem TE is from eq (6) Monograph/coordinate system reconsider2 re labelled as TE to identify energy between q=270 and q=90 or +90 and
-90
  Rem TE(n + 1, m + 1) = (Cos(L) * Cos(f(n)) * Cos(g) * Cos(pi / 4 + wa) - Cos(L) * Sin(f(n)) * Cos(g) * Cos(pi / 4 - wa) + Sin(L) * Sin(g) *
Cos(pi / 2 - wa)) * Cos(q(m)) + (Cos(L) * Cos(f(n)) * Cos(g) * Cos(-wa + pi / 4) + Cos(L) * Sin(f(n)) * Cos(g) * Cos(wa + pi / 4) + Sin(L) *
Sin(g) * Cos(pi - wa)) * Sin(q(m))
Rem TE(n + 1, m + 1) = ((Cos(L) * Cos(f(n)) * Cos(g) * Cos(wa) - Cos(L) * Sin(f(n)) * Sin(wa) + Sin(L) * Sin(g) * Cos(wa)) * Cos(q(m)) +
(Cos(L) * Cos(f(n)) * Sin(wa) * Cos(g) + Cos(L) * Sin(f(n)) * Cos(wa) + Sin(L) * Sin(wa) * Sin(g)) * Sin(q(m)))
TE(n + 1, m + 1) = ((Cos(l) * Cos(f(n)) * Cos(g) * Cos(wa) - Cos(l) * Sin(f(n)) * Sin(wa) + Sin(l) * Sin(g) * Cos(wa)) * Cos(q(m)) + (Cos(l) *
Cos(f(n)) * Sin(wa) * Cos(g) + Cos(l) * Sin(f(n)) * Cos(wa) + Sin(l) * Sin(wa) * Sin(g)) * Sin(q(m)))
 Rem Ez is from eq (12) Monograph/coordinate system reconsider2
  TEz(n + 2, m + 2) = (1 - z * Cos(q(m))) ^ 2 / (1 - z ^ 2) ^ 2 * TE(n + 1, m + 1)
  Rem TEz(n + 2, m + 2) = ((1 - z * Cos(q(m))) / (1 - z ^ 2)) ^ 2 * TE(n + 1, m + 1):
     If TEz(n + 2, m + 2) < 0# Or TEz(n + 2, m + 2) = 0# Then GoTo 9022 Else GoTo 9222
9022 Let flag = 0#: GoTo 9322
9222 flag = 1#
9322 TEzz#(n + 3, m + 3) = flag * TEz(n + 2, m + 2):
  Rem code below inserted to establish phase relationship between precession angle and orbital position of Earth
  Rem If m = 181 And wa > 270 / 180 * pi - stp And wa < 270 / 180 * pi + stp Then GoTo 9440 Else GoTo 3000
9440 Rem PSet (f(n), TEzz#(n + 3, m + 3)), QBColor(12)
STemp(n + 4, m + 3) = STemp(n + 3, m + 3) + TEzz#(n + 3, m + 3) * Cos(l):
If n = 359 Then GoTo 5022 Else GoTo 5822
5022 SSTemp(n + 4, m + 4) = SSTemp(n + 4, m + 3) + STemp(n + 4, m + 3):
5025 Rem SSSTemp(n + 5, m + 4) = SSSTemp(n + 5, m + 3) + SSTemp(n + 5, m + 3):
5822   If m = 364 Then GoTo 5422 Else GoTo 722
5422 If n < 359 Then GoTo 722 Else GoTo 5426
  Rem plot energy received between q= 90 to 270 and q=270 to 90 color it red
  Rem plot energy received as q goes from 270 to 90 degrees.
  Rem this plot is normalized to energy at w=270 and L=65
 Rem plot total energy received between 270 and 90
5426 If (SSTemp(n + 4, m + 4) / 7368.60821016331) < 1.0214 Then cg = 10 Else cg = 9
 D = 65520: DrawWidth = 3: PSet (stp * (v - 1), shf + (2.25 / D) * (SSTemp(n + 4, m + 4))), QBColor(cg):
 Rem if m>182 then q is between +90 and +270 blue

69

```
Rem to check this use code below'
  Rem Print "m="; m; " n="; n: stop
   Rem sum red and blue energy to obtain total energy received at a given precession angle,w.color it according to latitude.
    Rem PSet (stp * (v - 1), shf + (2.25 / D) * 1 / 2 * (SSTemp(359 + 4, 364 + 4) + SSEzz#(359 + 4, 182 + 4))), QBColor(L / pi * 18 - 5 / 10):
     Rem print data to determine slope at wa=180, the current precession angle
 If v = 52 Or v = 53 Or v = 54 Or v = 55 Or v = 56 Then GoTo 814 Else GoTo 722
814  FontSize = 10: dd = 0.022
ForeColor = QBColor(12): PSet (425 / 180 * pi, dd * (v) - 0.05 - l / pi * 2.53), QBColor(15): Print " "; Format((SSEzz#(n + 4, m - 182 + 4) /
7368.60821016331), "0#.####0"); " "; Format(SSEzz#(n + 4, m - 182 + 4) / 7368.60821016331 * 19.58 + 253, "###"); " "; Format(wa / pi *
180, "###"); " "; Format(l / pi * 180, "##")
933 ForeColor = QBColor(9): PSet (378.5 / 180 * pi, dd * v - 0.05 - l / pi * 2.53), QBColor(15): Print " "; Format((SSTemp(n + 4, m + 4) /
7368.60821016331), "0#.####0"); " "; Format(SSTemp(n + 4, m + 4) / 7368.60821016331 * 19.58 + 253, "###"):
Rem ForeColor = QBColor(0): PSet (200 / 180 * pi, dd * v - 0.05 - L / pi * 2.53), QBColor(15): Print " "; Format((1 / 2 * (SSTemp(n + 4, m
+ 4) + SSEzz#(n + 4, m - 182 + 4)) / 7368.60821016331), "0#.####0")
Rem If v = 55 Or v = 56 Or v = 57 Then GoTo 9000 Else GoTo 722
Rem 9000 PSet (0, 0.02 * v - 0.2): Print SSEzz#(4 + n, m - 182 + 4) / 7368.13799473677 - SSEzz#(4 + n, m - v - 182 + 4) / 7368.13799473677
Rem to find value of SSEzz# at w=180 set L=65/180*pi and insert line of code below.Read line above for value
 Rem PSet (0, v * 0.02 - 0.2): Print "m="; m; " wa="; (wa / pi * 180); " TS="; ((SSEzz#(n + 4, m - 182 + 4))); " "; Format(wa / pi * 180,
"0#.###"); (L / pi * 180):: Stop
722 Next n
73   Next m: v = v + 1
Rem U = U + 1
Next wa:Rem reset v for next L
v = 0
test = test + 1:
Next l:
 ForeColor = QBColor(0)
10 GoTo 6
PrintForm :Rem enter the name of the file
SavePicture Image, "C:\AvTEvswL20to80G.bmp"
6 End Sub
Private Sub Command2_Click()
End
End Sub
Private Sub Form_Load()
pi = 3.141592654: sc = 1: Scale (-0.1 * pi, sc)-(3 * pi, -sc): Top = 0: Left = 0:
Height = 12500:  Width = 1.3 * Height:
End Sub
```

www.ingramcontent.com/pod-product-compliance
Lightning Source LLC
Chambersburg PA
CBHW041450210326
41599CB00004B/204